Corporate Crops

Corporate Crops

Biotechnology, Agriculture,
and the Struggle for Control

GABRIELA PECHLANER

University of Texas Press ⏦ *Austin*

Support for this book comes from an endowment for environmental studies made possible by generous contributions from Richard C. Bartlett, Susan Aspinall Block, and the National Endowment for the Humanities.

Portions of chapters four and five previously appeared in the *Canadian Journal of Sociology* 35:2, pp. 243–269, and are reproduced by permission of the journal; portions of chapter six previously appeared in *Anthropologica* 52:2, pp. 291–304, and are reproduced by permission of the journal.

First edition, 2012

Library of Congress Cataloging-in-Publication Data

Pechlaner, Gabriela, 1968–
Corporate crops : biotechnology, agriculture, and the struggle for control / Gabriela Pechlaner. — 1st ed.
    p.   cm.
    Includes bibliographical references and index.
    ISBN 978-0-292-75687-8
    1. Monsanto Company—Trials, litigation, etc.   2. Agricultural biotechnology—United States.   3. Agricultural biotechnology—Canada.   4. Plant biotechnology—United States.   5. Plant biotechnology—Canada.   6. Transgenic plants—United States.   7. Transgenic plants—Canada.   8. Intellectual property—United States.   9. Intellectual property—Canada.   10. Family farms—United States.   11. Family farms—Canada.   12. Agricultural biotechnology—Law and legislation—United States.   13. Agricultural biotechnology—Law and legislation—Canada.   14. Plant biotechnology—Law and legislation—United States.   15. Plant biotechnology—Law and legislation—Canada.   I. Title.
SB106.B56P43   2013
630—dc23   2012021645

doi:10.7560/739451

First paperback edition, 2013

Contents

List of Acronyms

AAFC	Agriculture and Agri-food Canada
APAS	Agricultural Producers Association of Saskatchewan
APHIS	Animal and Plant Health Inspection Service (United States)
BRS	Biotechnology Regulatory Service
Bt	Bacillus thuringiensis
CBAC	Canadian Biotechnology Advisory Committee
CBAN	Canadian Biotechnology Action Network
CBD	Convention on Biological Diversity
CCC	Canola Council of Canada
CCGA	Canadian Canola Growers Association
CFA	Canadian Federation of Agriculture
CFIA	Canadian Food Inspection Agency
CFS	Center for Food Safety (United States)
CIPO	Canadian Intellectual Property Office
CWB	Canadian Wheat Board
EMPA	Environmental Management and Protection Act (Canada)
EPA	Environmental Protection Agency (United States)
EPO	European Patent Office
EU	European Union
FAO	Food and Agriculture Organization (United Nations)
FDA	Food and Drug Administration (United States)
GATT	General Agreement on Tariffs and Trade
GM	Genetically modified
GMO	Genetically modified organisms
GRAS	Generally recognized as safe
GURTs	Genetic use restricted technologies
IPR	Intellectual property rights

ITPGR	International Treaty on Plant Genetic Resources for Food and Agriculture
IUCN	International Union for the Conservation of Nature and Natural Resources
LL	Liberty Link
NAFTA	North American Free Trade Agreement
NASS	National Agricultural Statistics Service (United States)
NBS	National Biotechnology Strategy (Canada)
NFU	National Farmers Union (Canada)
NGO	Nongovernmental agency
NIH	National Institutes of Health (United States)
NRCC	National Research Council of Canada
OAPF	Organic Agricultural Protection Fund (Canada)
OCIA International	Organic Crop Improvement Association International
OIG	Office of the Inspector General (United States)
PBR Act	Plant Breeders' Rights Act (Canada)
PNT	Plants with novel traits
PPA	Plant Protection Act (United States)
PTO	Patent and Trademark Office (United States)
PUBPAT	Public Patent Foundation (United States)
PVP Act	Plant Variety Protection Act (United States)
rBST	Recombinant bovine somatotropin
RR	Roundup Ready
SAF	Saskatchewan Agriculture and Food
SAFRR	Saskatchewan Agriculture, Food and Rural Revitalization
SARM	Saskatchewan Association of Rural Municipalities
SCDC	Saskatchewan Canola Development Commission
SCGA	Saskatchewan Canola Growers Association
SOD	Saskatchewan Organic Directorate
SPS Agreement	Agreement on the Application of Sanitary and Phytosanitary Measures (World Trade Organization)
SSR	Seed Sector Review (Canada)
SWP	Saskatchewan Wheat Pool
TA	Technology Agreement (United States)
TBT Agreement	Agreement on Technical Barriers to Trade (World Trade Organization)

TNC	Transnational corporation
TRIPS Agreement	Agreement on Trade-Related Aspects of Intellectual Property Rights (World Trade Organization)
TUA	Technology Use Agreement (Canada)
UPOV	International Union for the Protection of New Varieties of Plants
USDA	United States Department of Agriculture
WTO	World Trade Organization

Acknowledgments

This book evolved from research I conducted for my PhD dissertation. My thanks and appreciation are therefore due to my PhD committee. Particular thanks are due to my senior supervisor, Gerardo Otero, who encouraged the production of this book. I would also like to thank Karl Froschauer and Murray Rutherford, the members of my dissertation committee, for their invaluable critique, suggestions, and timely questions. Thanks are also due to the Social Science and Humanities Research Council for the two years of doctoral funding that made my research trips possible.

This book would not have been possible without the many interviewees who agreed to talk with me and share some of their valuable time, sometimes during their busiest season. The people of Saskatchewan live up to their reputation of being kind and helpful people. In Mississippi, I was met with invaluable assistance not only from individuals, but from agricultural institutions as well. Special thanks are due to members of the Mississippi Farm Bureau and the Mississippi Agricultural Extension Service for putting me in contact with a number of farmers. Many people in both of these regions went far beyond giving their time, and generously extended their hospitality.

I would particularly like to acknowledge those involved in the litigation that forms the basis of my case studies. Speaking with anyone in the course of litigation is an exercise in trust, and I appreciate the efforts of those who were willing to engage with me in this way. In return, I have tried to present the positions of all those involved ethically and honestly. I hope I have been successful in doing so.

Last but not least, I would like to thank my family for their support during this long process. This book is dedicated in particular to my eldest, Selina, who suffered the brunt of my absence and distraction during its production.

Corporate Crops

Introduction

New technologies are not deployed in a historical vacuum. Rather, they are introduced into a particular set of social, economic, and ecological circumstances with established and knowable trajectories.
KLOPPENBURG, *FIRST THE SEED*

The Changing Face of Agriculture

In 1996, the first genetically modified crops were commercialized in North America. Adoption of these crops was subsequently rapid in both Canada and the United States, as was the proliferation of related litigation. Just two years later, in August of 1998, the Monsanto Company brought legal action against Saskatchewan canola farmer Percy Schmeiser, alleging patent infringement of its transgenic canola seeds. The resulting legal struggle splashed across the Canadian news when Schmeiser claimed that he had not deliberately obtained Monsanto's technology but had been involuntarily contaminated with it. Suddenly, the possibility of farmers being sued for patent infringement due to natural or involuntary processes became irrevocably associated with the new agricultural biotechnologies.

The issue of involuntary contamination was soon joined by a number of other concerns related to agricultural biotechnologies' proprietary aspects. The introduction of a plethora of new laws and contractual obligations associated with the technology—prohibitions on seed saving, restrictive technology agreements, even imbalanced incentive agreements and restrictions on herbicides—all suggest a significant and ongoing shift of what was previously under producer control to biotechnology developers, with an associated shift of economic benefit. Nor are these concerns strictly a Canadian phenomenon. In the United States, for example, the Monsanto Company has filed 112 lawsuits against farmers (Center for Food Safety [CFS], 2007). Genetically modified seeds, as patented, self-reproducing "inventions," have challenged both the traditional understanding of property rights and the historical dynamics of control over agricultural production more broadly.

The situation in Canada and the United States suggests that biotechnology is facilitating a social reorganization of agriculture whereby biotechnology corporations are using a variety of legal mechanisms associated with the introduction of biotechnologies to expropriate ownership and control over agricultural production from farmers. Further, the United States is a world leader with respect to the technology, and this reorganization appears to be enjoying global replication with the assistance of various international agreements, such as the World Trade Organization's Agreement on Trade-Related Aspects of Intellectual Property Rights. It would seem that a new regime of food production, based on corporate biotechnologies and what I call "expropriationism," is assured, with significant implications for global food security. While the trends are increasingly deeply set, however, there are indications that local-level activities—such as those occurring in the legal arena—can influence the terms of biotechnologies' adoption and usage.

The basis of the reorganization of agriculture lies with the granting of patent rights on life forms. Life forms, such as seeds, are self-reproducible. The granting of proprietary rights over such "inventions" lays ownership claim to untold varietal improvements from the selection efforts of plant breeders and farmers for generations preceding, by the addition of one patented trait. Further, it transfers this ownership of one of the most fundamental means of life—the ability to produce food—to a very concentrated sector. The result has been legal conflicts over seed saving and patent infringement, as farmers resist the conditions of the transition to biotechnology adoption. Intellectual property protection on plants is not new to agricultural biotechnologies, of course, but they facilitate a definitive step toward the further commodification of the seed.

Struggles over rights extend beyond the conflicts over seed saving brought by plant breeders' rights and patents, however. Genetic technologies can spread through the environment—even against the wishes of the owner of the property where they establish—through natural processes such as wind or animals, transfer on farm machinery or during transportation, or human error. With this potential and probable transfer of patented genetic material to lands not covered by legal contract, a whole new property rights issue is initiated. Patents placed on inventions that cannot be completely contained offer a direct challenge to the rights of farmers to the products of their land. While biotechnology companies do not want to relinquish control over their significant investments, an affirmation of ownership rights over self-reproducing inventions leaves farmers potentially liable for patent infringement over genetic material they may not want, or even be aware of, in their crops. Further, it challenges the rights of those—such as organic farmers—who may be harmed by it.

The resulting conflicts over agricultural biotechnologies pit farmers with limited access to resources against powerful and well-funded biotechnology companies. The exact extent of conflicts is difficult to determine. Only a limited number of infringement cases proceed to trial, and pre-court settlements typically require farmers to sign a nondisclosure agreement. But while the details of the Schmeiser case are contested, the significance of the issues is not. The emerging legal framework around agricultural biotechnologies raises many questions about the impact on farmers as well as broader issues of control over food production. The right to alternative agricultures and the imbalance of power between farmers and biotechnology companies are an integral component of such shifts in control. Even for those who embrace industrial agriculture, the plethora of rules and incentives that the Monsanto Company (in particular) has introduced in association with biotechnologies have increasingly constrained farmers to the extent that some allege that the company—which has an effective monopoly in some regions—has used them to create a virtual cartel. Ostensibly, the social utility of patenting is to stimulate private investment to accrue broader social benefits. The system needs to be considered as a whole in regard to such benefits.

The corporate structure of the new agricultural biotechnology industry itself raises important questions regarding the accelerated transfer of control over food production from more local and broad based to that which is supranational and concentrated. With biotechnology, inputs can now be fully integrated with production processes to produce end products tailored to corporate visions and needs for profitability. Corporate concentration thus means that production choices are further reduced. The top three seed companies—Monsanto, DuPont, and Syngenta—account for 47% of the global proprietary seed market. The top ten account for 67% (ETC Group, 2008:12). Proprietary ownership over germplasm and imbalances in farmer-industry relations have a significantly different meaning in the context of such concentration. This high level of corporate control could have the potential to render farmers, consumers, even nation-states increasingly irrelevant to agricultural production decisions.

For Schmeiser and those in his position, the legal tactics of biotechnology developers may seem to be nothing more than underhanded corporate bullying—the stripping of farmers' democratic and economic rights and an assault on national sovereignty over food production, all in the name of private profits. Less nefariously, these conflicts represent the inevitable tensions that result from any technologically induced production change, as the introduction of agricultural biotechnologies could herald the eventual corporatization of one of the last holdouts from capitalist processes of production. In either case, the introduction of agricultural biotechnologies has the potential to reorganize many of the social aspects of agriculture and to intensify the economic restructuring already in evidence. Due to the incredible speed with which agricultural biotechnologies are being adopted in the United States and Canada, the social implications of this direction of development are of considerable import, not only within these nations, but more broadly, given their global impact.

The Approach to the Problem

Biotechnology is broadly defined by the Convention on Biological Diversity as "any technological application that uses biological systems, living organisms, or derivatives thereof, to make or modify products or processes for specific use" (Food and Agriculture Organization [FAO], "FAO Statement"). This book, and the current controversies, revolves around what many call the "new biotechnologies": those technologies specifically involving genetic manipulation, as in the creation of transgenics (organisms altered by the introduction of genes from another species). Commercial application of genetically modified [GM] crops, or genetically modified organisms [GMOs], began in the mid-1990s. Since then, biotechnology adoption has increased at an astronomical rate.

From 1996 to 2010, biotechnology crop production area increased from 1.7 million hectares to 148 million hectares globally (James, 2010). By 2010 the technology's proponents touted its billionth hectare (when calculated cumulatively) (ibid.). Ninety-two percent of global agricultural biotechnology occurs in just six countries. The United States is by far the leader in agricultural biotechnologies, accounting for 45% (66.8 million hectares) of global production area in 2010, while Canada slipped from the third largest country involved in 2005 to occupy fifth (behind

Brazil, Argentina, and India) in 2010, with 6% (8.8 million hectares) of global production area. China, Paraguay, Pakistan, South Africa, and Uruguay are the only other countries with more than a million hectares, with the remaining 19 countries accounting for approximately only 2% of global crop area (percentages calculated from James, 2010). Canada and the United States are the only industrialized countries with significant GM crop area, and together account for 51% of global GM crop production. Indeed, prior to the explosion of adoption in developing countries, Canada and the United States commanded a much higher percentage. In 2000, for example, only four countries accounted for 99% of the 44.2 million hectares of global biotech production area. At that time, Canada and the United States accounted for 75% (69% and 7%, respectively) of the global production area (calculated from James, 2000).

While some innovative genetic modifications receive much public attention—vitamin A infused rice and drought-resistant cassava, for example—to date such innovations have either failed to reach commercialization or remain statistically negligible in contrast with the two key GM traits of herbicide tolerance and insect resistance. In 2010, GM crops were 61% herbicide tolerant, 17% insect resistant, and 22% stacked (James, 2010). Herbicide tolerance allows crops to survive the application of herbicide, thus allowing weed kill even after crops have emerged. The most common such applications are the Monsanto Company's Roundup Ready crops. Insect-resistant crops have been genetically modified to incorporate a pesticide, such as Bacillus thuringiensis, or Bt, into all cells of the plant to protect it from insects. These traits have been applied to a number of key agricultural crops—primarily canola, maize, soybeans, and cotton, and, most recently, beets and alfalfa. Minimal acreage is dedicated to a few other GM crops, such as squash and papaya.

The controversy over these crops has been heated. Popular media reports polarize those who view genetic modification in agriculture as the technological fix for hunger and nutrition-related disease against those who emphasize its negative environmental implications and the high level of uncertainty and risk. The "battle over biotechnology" has primarily highlighted biotechnologies' health and environmental risks, including such concerns as the health impact of GM foods, decreased genetic diversity, impacts on other species, and superweeds. Such a rapid adoption of any technology is likely to have significant social impacts as well, however. Social concerns are increasingly evident in the growing property and control issues associated with the technologies, such as the rising corporate control over food; the loss of farmers' traditional rights to save seed; decreasing production choices; a potential compromise of national food autonomy; biopiracy; and food security concerns, particularly for subsistence farmers.

This book's underlying focus is the issue of "choice." It is concerned with the extent to which social control can prevail under conditions of increasing capital investment and concentration in sectors (such as agriculture) that are dominated by the global investment strategies of transnational corporations. It focuses on the legal framework around biotechnology as the locus of struggles for control over agricultural production, and as a key component of a new corporate strategy for capital accumulation in agriculture both locally and globally. Its organizing principle is thus the question: To what extent is the introduction of corporate-driven proprietary agricultural biotechnology initiating a social reorganization of agricultural production, and to what extent is any such reorganization affecting the degree of control that Canadian and American farmers have over food production? I approach this question through four case studies of lawsuits such as Schmeiser's, conducted comparatively between the United States and Canada, the top biotechnology adopters among developed countries. To the extent that a degree of choice and social control around agriculture are retained, even the environmental and health issues associated with biotechnology have the potential to be resolved in a socially desirable way.

Despite the importance of these issues, even scholarly research on agricultural biotechnologies has overly emphasized biotechnologies' physical aspects. While it is indeed important to debunk the "technological fix" perspective of world hunger that is promulgated in the media (e.g., Boyens, 1999, 2001; Shiva, 2000b; Tokar, 2001), many of these arguments drift uncomfortably close to technological determinism, viewing the technology itself as inherently good or bad for society. More convincingly, it has been argued that the benefits of GM technologies are more likely to be determined socially, not scientifically (Kloppenburg, 2004, 2010; Lewontin, 2000; Middendorf et al., 2000). That is, the social benefits of the technology will be determined by the direction of its development and the uses to which it is put. For a simple example, agricultural biotechnologies can be used in the socially motivated development of drought-resistant crops or the profit-motivated development of genetic use restricted technologies [GURTs], more commonly known as "terminator" technologies (crops genetically modified to be nonreproducing).

Nonetheless, even the bulk of sociological research which touches on power relations related to agricultural biotechnologies has focused on its political or regulatory aspects, its social construction, and various groups' acceptance or resistance to it (for some examples, see Anderson, 2001; Andree, 2002; Bauer and Gaskell, 2002; Jones, 2000; Kinchy et al., 2008; Kurzer and Cooper, 2007; Lassen and Sandoe, 2009; Schurman and Kelso, 2003; Seifert, 2008; Walsh-Dilley, 2009). A mounting body of work addresses biotechnology in the context of North-South power differentials, including inequality issues in international trade agreements, the privatization of genetic resources, and the impact of capital-intensive technologies on low-income countries, among other issues (e.g., Arends-Kuenning and Makundi, 2000; Barton and Berger, 2001; Falcon and Fowler, 2002; Fitting, 2008; Gonsalves et al., 2007; Grace, 2002; McNally and Wheale, 1998; Otero and Pechlaner, 2005; Shiva, 2000a, 2000b, 2001). Aspects of power and control related to biotechnologies (and their associated proprietary issues) in industrialized countries have been largely neglected. While legal communities are abuzz with the new application of patent law to seeds (see, e.g., Hamilton, 2005; Mandel, 2004, 2005; McEowen, 2004; Park, 2003), there is a dearth of sociological analysis linking these proprietary changes in agriculture with empirical assessments of their impact, although interest along these lines (and with respect to power and control more broadly) is growing (see, e.g., Aoki, 2008; Busch et al., 1991; Kloppenburg, 2004, 2010; Kuyek, 2005; Mascarenhas and Busch, 2006; Mauro and McLachlan, 2008; Müller, 2008; Wield et al., 2010). In order to investigate these new dynamics, agricultural biotechnologies must be considered both locally and globally; their supportive regulatory structure, impacts, and resistances to them operate on both these levels.

The sociology and political economy of agriculture literatures have identified many trends in industrialization common to agriculture that can help us to interpret the local role of agricultural biotechnologies in agricultural production. Many of these trends demonstrate the piecemeal strategy to accumulation that has accompanied the natural limitations to the industrialization of agriculture (Goodman, 1991; Goodman and Watts, 1994). Scholars such as Goodman, Sorj, and Wilkinson (1987) have provided two conceptual tools—appropriationism and substitutionism—to distinguish these processes from those that occur in wholesale industrial transformations. While these terms emphasize capital accumulation strategies in the spheres of production and processing, there are indications that the whole package of proprietary changes associated with agricultural biotechnologies may be having impacts that these concepts cannot sufficiently explain.

Directly addressing this subject of technological change in agricultural crops, Kloppenburg's (2004) account of the political and economic history of the commodification of the seed clearly demonstrates that the introduction of any new technology will be shaped by existing social relations. Specifically, he argues that capital had two routes to pursue in attempts to commodify the seed: technical and social. Technical routes are found in such developments as hybrid seeds, whose second generation is not as effective as the first, motivating yearly purchase. With agricultural biotechnologies, the technological route is represented by terminator technologies, which do not regenerate at all (Jefferson et al., 1999). These have thus far been prevented from commercialization by widespread public protest.

Where technological routes are precluded, there is a social route, notably through intellectual property rights protections. In the United States and Canada, this was historically accomplished through such mechanisms as plant breeders' rights. With agricultural biotechnologies, this social route found expression in technology agreements and, finally, patents on germplasm. In conjunction with a number of other technology dissemination strategies, there is some indication that this social route is being taken not only as a means to commodify the seed, but also as a means of capital accumulation in itself. As noted by Kloppenburg, however, new technologies are not developed and disseminated in a vacuum. While there are powerful political and economic forces driving the development of these technologies, there are also forces of resistance, both within and outside the legal forum. These sociopolitical forces will need to be accounted for in any trajectory of the technology's development.

While the processes discussed here occur on the local level, there are strong arguments that globalization has shifted the locus of regulatory rule making outside of the national sphere. Some globalization scholars, for example, argue that as globalization has allowed capital to escape the bounds of the nation-state, these nation-states are now held hostage as transnational capital plays one state against another, endlessly maneuvering for the best comparative advantage (Strange, 1996, 2003; Teeple, 2000). Domestic restructuring along neoliberal free market lines is the necessary consequence. Further, subordinate classes, who once had limited access to state influence, have no such representation at the global level. Thus globalization arguably presents a serious disjuncture between supranational rule making and the potential for democratic input. In this scenario, even local-level resistance within the legal forum would be unable to influence global rules—such as those that apply to intellectual property rights—affecting the method of agricultural biotechnology dissemination.

These concerns are articulated in relation to agriculture and food by the food regime perspective, conceptualized by Harriet Friedmann and Philip McMichael (Friedmann 1992, 1993, 1995a, 1995b; McMichael, 1992, 2005; Friedman and McMichael, 1989). A food regime characterizes rule-governed relations of production and consumption on a global scale (Friedmann, 1993). The concept's proponents have identified two past regimes and characterize the current international structure of food production as similar to the "world car," whereby multinational corporations assemble processed foods from globally sourced components. These scholars further argue that the transnational restructuring initiated in the second food regime, in conjunction with the current wave of free-trade agreements, severely compromises the ability of nation-states to control their national agricultures. It is, therefore, very much a "neoliberal food regime," facilitated by capital-intensive technologies such as agricultural biotechnologies.

Thus, the features of the current period are characterized as particular agricultural manifestations of globalization tendencies more generally. In this neoliberal food regime, potentially mediating activities that occur within nations—such as through social movements or in the legal arena—are often under-considered. There is good reason to believe that they do mediate, however. A growing body of scholarship has suggested that local forces (including state policies, regional politics, and community resistance) can affect the implementation of the neoliberal globalism logic in agricultural production (see, e.g., Constance et al., 2003; Le Heron and Roche, 1995; Moran et al., 1996; Novek, 2003; Wells, 1997). Local responses to law are one of these forces. In fact, given that court rulings have the ability to trigger institutional change, the use of law may have the greatest potential for those opposed to biotechnologies' impact on agricultural production.

The Study

The first goal of this study is to investigate the extent to which the introduction of agricultural biotechnologies has initiated a social reorganization of agriculture, and how any such reorganization is affecting farmer control over agricultural production. Is there evidence that legal means are becoming a new capital accumulation strategy that expropriates farmers' control over their production? A subquestion to this study regards resistance in the neoliberal regulatory context: To what extent are local factors (local legal and community resistance) able to affect the trajectory of agricultural biotechnology development within the United States and Canada? Can such resistances still affect national regulation under the ideology of globalization? The focus for these questions is the proprietary and control aspects of the technology. An underlying concern is the extent to which the social reorganization of agriculture has broader social impacts, such as for environmental sustainability and global food security. While an investigation of these issues is beyond the present scope, the link to them is implicit. Corporations are driven by profit, not social goals, and consequently a transfer of control from farmers and society to a highly concentrated corporate sector will have some impact on these issues.

The research for this book was conducted using a case study approach (see Yin, 1982, 2003), structured comparatively between two regions within Canada and the United States, and based around selected GM seed–related litigation in each region. The selection of the regions themselves was determined by the location of litigation significant to the evolution of case law around the proprietary technology. The relevant lawsuits in Canada were self-evident. *Monsanto Canada Inc. v. Schmeiser* was the first—and as yet only—case of patent infringement litigation against a farmer for GM seeds to make its way through the Canadian court system. The case was groundbreaking not only in Canada, but also internationally, as a result of the issue of patent infringement in the face of involuntary possession. Following in its footsteps, and directly related to the issue of contamination, was another case originating in Saskatchewan, *Hoffman v. Monsanto Canada Inc.* Two organic farmers, Larry Hoffman and Dale Beaudoin, launched a class action lawsuit against Monsanto and Aventis (later Bayer) for the loss of their organic canola market due to contamination with GMOs. This was the first incidence of farmers going on the offensive in an attempt to impose liability on biotechnology developers for the technology's impact.

In the United States there is significantly more legal action between farmers and biotechnology companies, although, until very recently, no case directly addressed the issue of involuntary contamination or corporate liability. Two lawsuits in Mississippi—the Scruggs case (*Monsanto Co. v. Scruggs*) and the McFarling case (*Monsanto Co. v. McFarling*)—stood out for their incorporation of broader control issues into their patent infringement disputes. Monsanto launched these two suits against the farmers for saving seeds with Monsanto's patented technology. Through their defense, both cases acted as a direct attack on the new regime by challenging the validity of patents on plants, and—particularly in the Scruggs case—challenging the overall structure of technology dissemination. Mississippi also has a disproportionately high number of such patent infringement cases, further supporting it as a region of interest.

Each case study involves an investigation of the intellectual property context and the changes brought by the selected lawsuits (assessed through court decisions, related court documents, and interviews with litigants and their legal representatives), as well as broader interviews with agricultural stakeholders on the changes brought by the proprietary technology. Approximately 80 semistructured and primarily face-to-face interviews were conducted in Mississippi and Saskatchewan (40 in each region) over the spring and summer of 2005 (see appendix for details). Through these interviews, I investigate how biotechnology's introduction and its accompanying legal framework affect farmers: Are they being economically "bullied," as some claim, by their inability to match corporate dollars on the legal front? Are they making production choices (e.g., GM crop adoption) based on a desire to avoid legal "double-binds"? Are farmers losing control over traditional rights (such as the right to save seeds) for the sake of corporate patents, while facing a compromise of their own ownership rights (e.g., through contamination of organic crops)? How are farmers reacting to any perceived negative social impacts resulting from biotechnologies?

While the evolving body of legal precedents and regulations provides insight into the legislative framework unfolding around biotechnology, the interviews help get beyond the "law on the books" to reveal the needs, perceptions, and, ultimately, application of this framework—the "law in action" (Greenbaum and Wellington, 2002; Sumner, 1979). As these legal precedents are still unfolding, many of the repercussions will only be evident further down the road. Widespread concern over the issues, preventative actions, and disjuncture between perceptions and the current state of jurisprudence, for example, are significant qualifiers of the "law in action" and can assist in predictions of the future impact and of resistance to this impact.

Organization of the Text

This book first situates agricultural biotechnology in its theoretical and regulatory context and then moves on to provide empirical evidence of the reorganization of agriculture and opposition to it in Saskatchewan and Mississippi. Chapter 1 draws heavily on political economy of agriculture literature to contextualize the changes that proprietary agricultural biotechnologies have brought to agricultural production. I argue that the advent of agricultural biotechnologies is actually resulting in a new form of capital accumulation in agriculture—for which I propose the term "expropriationism"—based on these proprietary aspects. Further, it draws on the insights from globalization and food regime literatures to question whether it is indeed the case that the supranational context of agricultural production and dissemination results in a decline in state autonomy and regulatory control over agricultural biotechnologies. I argue that local-level activities—including general acts of resistance to the technologies and more specific acts in the legal forum—can affect a nation's role in the food regime and thus the shape of the regime itself.

Chapter 2 provides the regulatory context of the technology in Canada and the United States. It argues that Canada has played tag-along development to the U.S. biotechnology industry. While slightly more hesitant to patent life, Canada's regulations are ultimately evolving similarly to those in the United States, and the two countries appear to be aligning in a pro-biotechnology global bloc, emphasizing strong intellectual property rights and weak regulatory oversight. The European Union, discussed in contrast, provides insight into the international contestation over agricultural biotechnologies, which can influence the shape of the unfolding regime.

The next five chapters focus on the empirical research. The Canadian case is first, chronologically, largely because complex issues raised by the proprietary aspects of biotechnology have found greater purchase there in public discussions. This is in part because litigation in Saskatchewan highlighted the issue of involuntary contamination, whereas in Mississippi it unequivocally focused on patent infringement due to seed saving.

Chapter 3 documents Saskatchewan farmers' experience with the technology. Interviews reveal that the majority of producers are supportive of the technology because of the time and efficiency gains it provides, and while the proprietary aspects of the technology—specifically Monsanto's technology agreement[1]—were an irritant to some, the presence of alternative varieties and crops mitigated this irritation somewhat. Chapter 4 continues the Saskatchewan case study with a specific focus on local-level resistance to the technology triggered by the attempted introduction of unwanted Roundup Ready wheat and the relatively high population of organic farmers in the province. There is strong evidence to suggest this resistance has affected the industry's development in the region. These dynamics are virtually absent in Mississippi. Chapter 5 focuses on the selected litigation in Saskatchewan. In these cases, we see that in the clash between the property rights of farmers and the patent rights of technology developers, the trend has been supportive of the rights of the latter, with technology developers gaining some of the most important benefits of ownership while remaining exempt from its liabilities.

Chapters 6 and 7 present the Mississippi case. In Chapter 6, we see that Mississippi farmers demonstrated much greater extremes in their relationship toward the technology than did Saskatchewan farmers. Greater weed and insect pressures in Mississippi contribute to an appreciation of the technology that borders on technological dependence. At the same time, there is a distinct sentiment that Monsanto wields a monopoly control that is leaving farmers powerless in the face of its dictates. Interestingly, while there is less expressed support for those involved in litigation than in Saskatchewan, the issues raised in court have a great deal of resonance with the concerns of Mississippi farmers more broadly, especially in terms of patent misuse and antitrust. In Chapter 7, we see a trend that again indicates that case law is evolving in a manner that discounts these concerns and upholds the rights of patent holders.

Chapter 8, the conclusion, argues that these cases provide clear indications that the rise of proprietary technologies introduces a new means of capital accumulation, and that this rise is decreasing farmers' control over agricultural production. While we are still in the early stages of determining how the clash of rights raised by agricultural biotechnologies will ultimately be resolved, the case studies here are very supportive of a conclusion that the legal strategies associated with agricultural biotechnologies have indeed facilitated a new form of capital accumulation in agriculture. I also argue, however, that these trends are not globally inevitable. Resistance to these proprietary and control changes is occurring in both the legal forum and in the broader community. Litigation has become a particularly important locus of struggle over these issues in Canada and the United States. To date, the impact of this resistance has been piecemeal. But biotechnologies' role in the evolving international regime of food production does remain subject to local forces—a factor of some importance as proponents bolster the technologies' advancement by adding climate change to global hunger as candidates for its technological fix.

CHAPTER 1

Agricultural Biotechnologies on the Farm and around the World

Introduction

While agricultural biotechnologies have brought a significant number of changes to agricultural production, such technologically induced change is not new to agriculture, of course, and is well documented in scholarly literature. Biotechnologies' proprietary aspects add a new component to such change, however, which have the potential to instigate a social reorganization of agricultural production. Moreover, they may even be introducing a new capital accumulation strategy. Further, it is simply not possible to assess national agricultural change without attention to the global context, and their reciprocal influence. Even historically speaking, agriculture in North America has developed squarely in the context of international trade relations. The current era of "globalization" suggests a deepening of global integration, which lends even greater significance to the international context of national agricultures. At the same time, to assume that local activities are irrelevant in the shaping of the global food regime is to greatly undervalue the myriad influences on the national and subnational level: these influences form the legal and regulatory basis from which new technologies—such as agricultural biotechnologies—are nurtured and can develop into international forces . . . or fade away completely.

This chapter will first investigate relevant sociology and political economy of agriculture literatures for what they can contribute to an analysis of capital accumulation strategies in agriculture. It outlines how these literatures apply to GM technologies and where they fall short in explaining developments related to the new technology. Specifically, I suggest that two theoretical concepts of agricultural industrialization identified by Goodman, Sorj, and Wilkinson (1987)—appropriationism and substitutionism—need to be joined by a third, which I term "expropriationism."

Second, this chapter will seek to place agriculture in its global context by looking at theories about globalization more broadly, relating these to agriculture and food using the food regime perspective. The most convincing approaches to globalization incorporate factors that affect differential integration into the current wave of global economic restructuring. Similarly, the food regime, like globalization itself, is a contested project, vulnerable to resistances. What resistance might be significant enough to alter the shape of the socially problematic third food regime? I argue that local resistance to the technologies' expropriationist tendencies could be a strong enough reciprocal force to reflect upward on globalization tendencies and change the face of the third food regime.

Down on the Farm: The Industrialization of Agriculture?

From the classics to contemporary scholarship, political economy of agriculture literature provides many insights into historical capital accumulation trends in agriculture (see, e.g., Berlan, 1991; Buttel and LaRamee, 1991; Friedland, 2002; Friedmann, 1995b; Kautsky, 1988 [1899]; Lenin, 1964 [1899]; Thompson and Cowan, 1995). Many of these literatures have highlighted historical trends of industrialization that are applicable to agriculture: increased capitalization, concentration of agricultural input suppliers and output purchasers; substitution of independent producers with agribusinesses; increased productivity; the externalization of environmental costs; and the transformation of consumption patterns, among others. In some cases, the parallels with industrialization are drawn to the extent of rejecting agriculture's analytical separation from industry (Goodman and Watts, 1994:3). Despite the similarities, agriculture has retained a number of distinctions from industry due to its particularities as a nature-based process. Many aspects of agriculture deviate from typical capital accumulation patterns, and consequently theoretical distinctions are required.

Like Goodman and Watts (1994), I argue that the natural processes of agriculture do in fact render it exceptional to industrialization. Scholarly works that account for rather than artificially downplay this exceptionalism provide the greatest insight into agriculture, and the greatest predictive capacity for its future development. In particular, the conceptual tools of appropriationism and substitutionism, developed by Goodman,

Sorj, and Wilkinson (1987), provide an analytical framework through which many historical as well as current developments in agriculture can be viewed. As noted, developments in biotechnology are introducing new forums for capital to meet agriculture that cannot be accounted for by these concepts. Notably, a network of legal obligations associated with the technologies suggests that legal means may have joined these traditional capital accumulation strategies in agriculture.

Further, legal altercations over the issue of infringement and involuntary contamination highlight growing issues with respect to the genetic ownership that directly pit the rights of farmers against the rights of industry. Biotechnology critics additionally claim that contamination issues legally intimidate farmers from continuing production with anything other than GM seed stock. Consequently, while supportive of the concepts developed by Goodman et al., this section suggests that a new concept needs to be added in order to account for these legal changes, while retaining the ability to articulate agriculture's exceptional development. I propose the term "expropriationism."

Conceptual Tools in the Political Economy of Agriculture

While early political economy of agriculture literatures found some resonance of the experience of agriculture with broader theories of industrialization, it was not without some theoretical cost. Goodman, Sorj, and Wilkinson (1987) state that classical attempts at theorizing agriculture's position in capitalist development have resulted in conceptual distortions and debates inappropriately focused on social relations of production or the relative benefits of peasant versus capitalist modes of production (145). Such attempts at draping agriculture in the conceptual cloak of industrialization, they argue, overlook the central problematic of agriculture in capitalist development: its status as a natural process. Where agriculture's natural aspects produce impediments to wholesale industrial transformation, capitalist development finds other ways of incorporating agriculture into its processes—notably, through the piecemeal incorporation of discrete production elements.

The attempts to draw agriculture into broader theories of industrial restructuring did not stop with classical approaches. Goodman and Watts (1994) identify the concept of "Fordist agriculture" as one such attempt, whereby political economy and regulation theories have tried to overstate agriculture's industrialization in an effort to reject its exceptionalism from industry. Goodman and Watts argue that the Fordist agriculture conceptualization fails under empirical assessment, however. While aspects of the processing and input sectors of agriculture may demonstrate Fordist tendencies (e.g., high volume, standardized production and consumption), the conceptualization cannot be sustained with respect to labor at the point of production or to regulation. The significance of such conceptual slippage is not minor. Rather, Goodman and Watts argue that washing over agriculture with the "gloss of Fordism" overlooks important exceptions that need to be explained and consequently distorts a significant analytical question: "How does the organization of agricultural production and rural space change under different regimes of accumulation and modes of social regulation?" (15). This way of thinking puts agriculture firmly under the umbrella of its broader political economic context, but without creating a forced marriage of industrialization concepts and empirical evidence.

With a similar eye to analytical specificity, Lewontin (2000) argues that classical capitalist concentration failed in farming because of the sector's financial and physical particularities: the ownership of farmland is financially unattractive; labor is hard to control because farms are spatially extensive; economies of scale are limited; it is largely impossible to reduce the reproduction cycle; and the risks involved in farming—such as from weather, disease, and pests—made direct ownership in agriculture unattractive to capital (95). As a result, capital concentrated on the farm inputs and processing sectors in order to capture profits:

> The problem for industrial capital, then, has been to wrest control of the choices from the farmers, forcing them into a farming process that uses a package of inputs of maximum value to the producers of those inputs, and tailoring the nature of farm products to match the demands of a few major purchasers of farm outputs who have the power to determine the price paid. Whatever production risks remain are, of course, retained by the farmer. (96)

The concepts of appropriationism and substitutionism, developed by Goodman, Sorj, and Wilkinson (1987), provide a means of accounting for this piecemeal approach to capital accumulation. While their book is now somewhat empirically dated, the concepts hold their explanatory value for many processes in agriculture today, an indication of the usefulness of this kind of conceptualization. The two concepts overcome the aforementioned theoretical errors precisely because they focus on the way in which agriculture is exempted from traditional industrialization. Goodman et al. argue that agriculture, as productivity rooted in the natural processes of the earth, could not be brought wholesale under the control of capital due to the natural limitations of land, time (plant and gestation cycles), and biological processes (photosynthesis). Some of these limitations may be reduced—most particularly with the advent of agricultural biotechnologies—but, to date, capital has had to find other means of infiltrating the sector. In response to these natural barriers, capital has pursued a piecemeal and discontinuous path of agricultural industrialization through appropriationism and substitutionism.

"Appropriationism" is the "discontinuous but persistent undermining of discrete elements of the agricultural production process, their transformation into industrial activities, and their re-incorporation into agriculture as inputs" (1987:2). By definition, appropriationism occurs in the production sphere of agriculture, where competitive industrial capitals "create sectors of accumulation by re-structuring the inherited 'pre-industrial' rural production process" (8). The trajectories of appropriation therefore depend on the particular history already in place. In nineteenth-century Britain, for example, limited land and plentiful labor led to accumulation strategies based on "high farming" (the replacement of farm-produced animal fodder and organic manure by purchased oilseed cake and fertilizers [28]). In the United States, on the other hand, land was plentiful, and early appropriationism there focused on mechanization and automotive engineering (e.g., replacing horses and labor with tractors). Goodman et al. note that as the agronomic problems of mechanization and extensive cultivation later became evident in the United States, these trajectories merged. Further accounts of such appropriationist processes of accumulation are provided in great detail by Goodman et al. and others (e.g., Berlan [1991] on the "power age" and Kloppenburg [2004] on hybrid technologies).

"Substitutionism" follows a similar process by replacing agricultural products with industrial ones. While appropriationism seeks to advance capital accumulation in all facets that can be replaced in agricultural production, substitutionism seeks to replace the agricultural end-products, reducing them to industrial inputs for manufactured products. Thus, substitutionism mainly occurs in the processing of agricultural products and seeks to "interpose mechanized industrial processing and manufacture between the source of field production and final consumption" (Goodman et al., 1987:60). The first wave of substitutions resulted from mechanical processes of adding value (e.g., flour milling). Preservation technologies, such as canning and refrigeration, provided another avenue.

The advances in distribution capabilities that these technologies brought facilitated the international division of labor and the vertical integration of capital. Goodman et al. suggest that the production of margarine heralded a qualitatively new form of substitutionism: that of "industrial substitution as product creation" (69). Margarine broke the tight association between agricultural product and end-product by using cheap industrial raw materials to create a fully industrial substitute for an agricultural product. In consequence, agricultural products "assume[d] the status of industrial inputs, being used interchangeably as determined by cost and technical criteria" (69). From this point on, the power of substitution in accumulation strategies only expanded.

Noteworthy for present purposes is the historically and naturally contingent process of capital accumulation that appropriationism and substitutionism characterize. Scientific and technological developments are key factors for these processes as they provide new opportunities for capital advancement. State policy and the manner in which capital adjusts to it are also central to these accumulation strategies. For example, the U.S. government's policy of institutionalizing production surpluses (discussed further below), which relegated market forces around grains "to a secondary role," ultimately founded a new appropriation strategy as cheap grains supported the expansion of the "livestock feed and fattening industries" (13–14). The state is also central to accumulation strategies, as it is both an essential backer and regulator of technological innovations, such as agricultural biotechnologies. These local-level processes reveal the dynamic nature of the resulting agro-industrial complex, in which capital responds to the intersection of history, state policy, and advances in science and technology. Ultimately, these accumulation strategies have functioned to minimize the economic significance of agricultural production and reduce the power of farmers, who are sandwiched between successful accumulation strategies in the input and output sectors.

With the introduction of agricultural biotechnologies, the opportunities for appropriationism and substitutionism in agriculture are greatly expanded. Goodman et al. argue that biotechnology may indeed even herald a new epoch in agricultural appropriation strategies. While industrialization processes have historically worked to "diminish the uncertainties of nature" by rendering the elements of agricultural production "increasingly measurable and predictable" (120), biotechnologies provide opportunities to bring nature even further under industrial control. We already see evidence of this in the two main GM traits, of which bacillus thuringiensis [Bt] and Roundup Ready [RR] technologies are cases in point. Prior to the introduction of Bt cotton, for example, farmers needed to check their fields regularly for signs of bollworm infestation and to act quickly in the case of its occurrence. The use of Bt cotton prevents such an infestation, thus reducing the need for this labor (in exchange for a significant input expense). Similarly, while conventional weed control involves a number of steps, including tilling and repeated, careful, herbicide applications, RR crops reduce this to a quick application of Roundup over the growing crop.

Substitutionism is another venue where biotechnologies appear to move food production further under the auspices of bio-industrialization. Early substitutionism conducted using the techniques of chemical engineering can now occur with even greater separation from natural processes with the assistance of industrial microbiology (123). The dramatic changes possible in the food industry from the use of biocatalysts were already demonstrated in the case of high-fructose corn syrup, developed through the process of mutagenesis. The process made possible the substitution of corn (subsidized by surplus policies in the United States), and eventually other grains, for cane sugars as sweetener inputs for the processing industry. The result was a dramatic loss for tropical countries (Friedmann, 1992). Genetic modification provides the means to further improve such biological catalysts and to vastly increase the ability to disaggregate agricultural products into variously substitutable components. Agriculture is thus transformed from the production of crops—such as cotton, potatoes, corn, beans—into the production of inputs—fibers, starches, sugars, oils—for the food processing sector that are highly exchangeable and that can be globally sourced and resourced, according to industry dictates. Genetically modified vanilla as a substitute for natural vanilla, for example, presents a significant threat to farmers in a limited number of countries that produce the high-value crop, such as Madagascar (Suchitra and Surendaranath, 2004). Similar research is under way to create substitutes for other developing-country crops such as coffee and cocoa.

While never completely distinct processes, biotechnologies provide significant opportunity for heightened convergence of the processes of appropriationism and substitutionism since crops can be genetically modified with specific traits desired by processors. Over a decade ago, Friedmann argued that if the dominant tendencies were to reach their logical end, "farms would adapt production to demand for raw materials by a small set of transnational corporations . . . and in order to meet quality standards would buy inputs and services from (often the same) transnational corporations" (1992:379). This linking is already evident in contract farming, which is common in poultry production, for example. In the highly vertically integrated poultry sector, farmers are contracted by a company which supplies all the inputs (the feed, chicks, and veterinary supplies), requires compliance with specific management procedures, and procures and markets the final product. As agricultural biotechnologies move from the first-generation focus on production traits to the second-generation focus on consumer benefits (vitamin- or antioxidant-enhanced "functional foods," for example), food processors will be even more interested in such linkages.

One consequence of the increased linkage between inputs and end-products is that it furthers the corporate concentration already evident in the sectors that bookend agricultural production, and thus decreases farmers' power between the two while consolidating power over the food supply into an increasingly limited number of hands (see, e.g., Hendrickson and Heffernan, 2007). Of particular relevance here is that as this concentration increases in the inputs sector, the potential for alternatives — with respect to input suppliers and to non-GM varieties — decreases according to the dictates of these suppliers. In 2007, the top ten seed companies accounted for 67% of the global proprietary (patented) seed market (ETC Group, 2008:11). Eighty-three percent of the commercial seed market is proprietary seed (Context Network, cited in ibid.:11). The top ten agrochemical companies control 89% of the global market (ibid.:15). Given the linkages between these sectors, many of these are the same companies. For example, globally, Monsanto controls 23% of the proprietary seed market and 9% of the agrochemicals market. Indeed, according to Middendorf et al. (2000), every major seed company has some form of direct link to a chemical company.

Concentration in GM seed is even higher. In 2004, the Monsanto Company accounted for 88% of the global GM crop area. The company's global market share in key GM crops is also extremely high: 91% in soybeans, 97% in maize, 64% in cotton, and 59% in canola (ETC Group, 2005a, with statistics compiled from ISAAA and Monsanto). In 2005, the Monsanto Company purchased Seminis, a vegetable seed company, and became the world's largest seed company and instant market leader in global vegetable seeds (for example, Monsanto's global market share is now 31% in beans, 38% in cucumbers, and 34% in hot peppers [ibid.]). One can only speculate that a push in GM vegetable seeds is to follow. The high capital and long time frame involved in biotechnology development makes it a high-stakes investment and keeps the number of competitors low. Nonetheless, while many biotechnology developers are losing money, a few are making large profits. In 2004, the top ten biotechnology companies accounted for 72% of revenues (but only 14% of research and development) (ETC Group, 2005b). The Monsanto Company, a litigant in the lawsuits considered here, is one of these companies. Its posted second quarter net income was $543 million in 2007; it was $440 million for the same period in 2006 (Gillam, 2007).

While biotechnology provides countless means to extend traditional capital accumulation strategies in agriculture, the technology provides a major breakthrough in privatization strategies around the seed, something that has previously faced many historical impediments. Thus the privatization of germplasm provides an effective exemplar of capital's struggle to accumulate in agriculture—given the natural limitations of the seed's reproducibility—and of biotechnology's seemingly decisive role in this struggle. Jack Kloppenburg's seminal *First the Seed* (2004) outlines the capital accumulation patterns that historically evolved in the United States as a result of this natural limitation to full commodification. He asks: "Have plant breeding and seed production become a means of capital accumulation? If so, how has this been accomplished and what have been its effects?" (8).

Kloppenburg demonstrates that technology choice is highly dependent on the economic interests involved. He argues that accumulation in U.S. seed production occurred through two (often overlapping) routes — one social and one technical. Social routes include legislation designed to protect plant breeders, such as the commercial protection of plant matter afforded under Canada's Plant Breeders' Rights Act [PBR Act] or the United States' Plant Variety Protection Act [PVP Act]. A technical solution for some crops arrived with the advent of hybrid technologies (which prevent commercial-grade regeneration) in the 1930s. According to Kloppenburg, varietal improvement through hybridization instead of through open-pollinated varieties (which could be replanted) was a social, not a technological, choice. It was driven by agitation from the seed industry and is testament to the capacity of private interests to affect the direction of technological advances. Public-sector development and disbursement of seed varieties were also impediments to the commercial seed industry. In the United States, the privatization trend included a retrenchment of such activities and a progressive reorganization of research for commercial purposes, such as by relegating the public sector to basic research, which the private sector then applies and commercializes. A similar reorganization is now occurring with Canada's agricultural biotechnologies, as we will see.

Despite these efforts, hybrid technologies and plant-breeding legislation have provided only a partial solution to capital accumulation in agriculture. Biotechnology now allows for far greater gains. Ultimately, in a bid to finally end the impediments to accumulation in germplasm, the biotechnology industry has produced patented germplasm and terminator technology. Both routes require the farmer to purchase seed afresh each planting season, the former by physically precluding regeneration, and the latter by legally doing so. Terminator technology has already been retracted once due to public outcry over its risks to food security, particularly in developing countries (Vidal, 1999). The technology highlights concerns already raised by the development of biotechnologies according to profit rather than social dictates. Opportunities for socially beneficial agricultural biotechnologies abound—for example, drought-resistant or salinity-tolerant crops for food-insecure regions—but thus far corporate biotechnology has emphasized GM developments that can produce the greatest profit. Consequently the emphasis has been on high-intensity chemical farming of a limited number of mono-cropped varieties for regions already historically producing surpluses. In short, "what is profitable is not always coterminous with what is socially optimal" (Kloppenburg, 2004:150). While biotechnologies' physical aspects certainly raise some issues, the focus here is on their proprietary aspects.

The proprietary aspects of biotechnology are multifaceted and ethically complex, and cover such issues as the morality of patents on life, the social significance of seed saving, global equity, and shifting property rights, to name a few. The granting of general utility patents on plants and/or components of plants was not a forgone conclusion, but was again a social decision that required extensive industry effort and a supportive state environment. Eventually, a corporate-friendly proprietary framework was accomplished for biotechnology:

> By 1994, within 21 years of the advent of microgenetic engineering technology, the "bio-industrial complex" had achieved the categorization of biotechnological products and processes within the realms of the patentable at both the US PTO [Patent and Trademark Office] and the EPO [European Patent Office], had persuaded the European Commission to draft a Directive on biotechnological patenting in the Community, and had laid the foundations for the globalization of intellectual property

rights through the GATT [General Agreement on Tariffs and Trade] and the UN Convention on Biological Diversity, with the objective of securing worldwide patent protection for the products and processes of modern biotechnology. (McNalley and Wheale, 1998:310)

As GM seeds are produced and sold as patented inventions, farmers must obtain their seed commercially for every planting. The control over the food supply that this legal fact grants to corporations is a source of much contention. While proponents of the technology state that those who object can simply respond by not purchasing the technology, opponents claim that conventional (non-GM) seeds are increasingly difficult to obtain and that food security issues should not be decided by the economic imperatives of individual farmers. Even more significantly, this emphasis on producer choice denies the reality of the technological treadmill. Further, concerns over corporate control of the seed supply by legal fiat become extremely problematic when reproduced on a global scale, reaching the areas of many of the world's poorest farmers.

Although state support of the intellectual property rights of the biotechnology complex may be a calculated geopolitical move in countries such as the United States and Canada, as will be discussed, it has nonetheless opened the door to significant clashes of rights between farmers and biotechnology developers. While GM crops may be patented inventions, they are self-reproducing patented inventions which can spread through the environment—even against the wishes of the owner of the property where they establish themselves. This potential, even probable, transfer of patented genetic material to lands not under patent contract initiates a whole new form of property rights conflict. Farmers, who traditionally have had the right to the products of their land, can find themselves in direct conflict with biotechnology companies, who claim the right to their patented invention, wherever it ends up.

The result of this clash of rights initiated by the self-reproducing patented technology is a seemingly endless series of questions. With respect to involuntary GM presence, who owns the resulting progeny? Can farmers lose their crop, or be held liable for patent infringement for involuntary contamination? How much GM presence makes a crop subject to patent infringement? In patent infringement lawsuits, is justice possible in a context of extreme economic imbalance? Are pre-court settlement agreements fair, or are they based on farmers' inability to match corporate dollars on the legal front? If biotechnology developers can extend their ownership rights to succeeding generations, is this ownership associated with any of the traditional responsibilities of ownership? Are they responsible for removing unwanted spread of their technology? Are they liable if their technology creates other negative impacts, such as contaminating organic producers?

In sum, these questions ask whether farmers are facing a loss of control over traditional production rights (such as the right to save seed and the right to the products of their land) for the sake of biotechnology developers' intellectual property rights. At the same time, they ask whether any potential gain in these developers' rights is associated with the traditional responsibilities of ownership (e.g., responsibility for contamination). Whether farmers are subject to a loss on both fronts is still being determined. The legal resolution of these questions is unfolding in litigation in the United States and Canada. Ultimately, these questions suggest that a new form of capital accumulation may be occurring. Further, this new form of accumulation may itself expedite the transition to GM crops through legally supported genetic occupation, economic intimidation, and the closing off of production alternatives. While resolution on many of these issues is still pending, sufficient activity has occurred to indicate the trajectory of that resolution, and to consider whether there is sufficient evidence that this indeed represents a new form of accumulation in agriculture.

As stated in the introduction, the organizing question for this research asked: To what extent is the introduction of corporate-driven proprietary agricultural biotechnology initiating a social reorganization of agricultural production, and to what extent is any such reorganization affecting the degree of control Canadian and American farmers (and by extension, society more broadly) have over agricultural production? I suggest that a new capital accumulation strategy—one I term "expropriationism"—may be acting to further reduce farmer and societal control over agricultural production.

While appropriationism and substitutionism emphasize accumulation strategies occurring in the spheres of production and processing, expropriationism is proposed as an accumulation strategy that occurs through the network of legal mechanisms associated with agricultural biotechnologies. I define expropriationism broadly in order to capture a range of legal strategies for capital accumulation in agriculture associated with the introduction of biotechnologies. While patents on seeds comprise a key component of expropriationism, the concept actually describes an assemblage of legal mechanisms used in concert to shift the relationship between technology producers and developers in such a way that they restrict the power of farmers and facilitate a new capital accumulation strategy. The use of social restrictions as a means of accumulation in agriculture is not new: plant breeders' rights, producer contracts, and even a limited number of plant patents predate the introduction of biotechnologies. What is new, however, is the widespread introduction of an assortment of legal mechanisms associated with a specific technology that is itself becoming widespread, at least in the commercial production of some key crops. Thus expropriationism is defined less by an individual legal mechanism than by a convergence of mechanisms, including patents, technology agreements, incentive agreements, and various rules associated with biotechnologies. It suggests "a new form of capital accumulation that is bound up with the seed, but actually transcends it, as capital is extracted not just through the seed, but through new systems of power and control associated with its purchase and use" (Pechlaner, 2010:293).[1] Evidence of these changes can be seen in law, as the technologies' proprietary framework evolves through litigation, and in practice, through the changing relationship of farmers to their production system.

The term "expropriationism" differs from its conventional legal and Marxist usage of expropriation conducted by a public body ostensibly for public good. The expropriation occurring here is not for public benefit—arguments regarding the public utility of promoting private accumulation for technological advancement notwithstanding—but is in keeping with the neoliberal trend of accumulation through dispossession (Harvey, 2003). To a limited extent, this is based in an ideologically motivated position on public benefit. More conventionally consistent with the above terminology, however, is that if not directly employed by a state body, the strategy is certainly state facilitated. In short, the available avenues for capital accumulation are highly dependent on a number of historical and natural conditions, technological developments, and state policies. As in appropriationism and substitutionism, the state plays a key role in the support of this accumulation strategy, as will be discussed in more depth presently.

Given the high degree of contestation around the technology, the possibility remains that the state could change its pro-biotechnology position and regulate the industry according to more social dictates. The significant and abrupt overstepping of farmers' rights implied by biotechnology-related litigation has provided a catalyst for even broader civil society response, and the resulting social movement agitation and lobbying efforts around GMOs have intensified the pressure on governments. However, there is no shortage of arguments that globalized capital and capital-friendly transnational agreements have compromised national regulatory ability to the point where nations can no longer control their national interests even when motivated to do so. Ironically, then, just when opposition to the technology seems to be reaching its head in many developed countries, it may nonetheless be unable to affect any significant change. If globalization renders nation-states incapable of independent regulation, then nation-based struggles over biotechnology are immaterial, and the corporate regime of accumulation will prevail.

Globalization and the Food Regime Perspective: The Decline of the Nation-State?

Whatever else can be asserted about the concept of globalization, it has garnered an indisputable amount of both popular and academic attention: it is alternately feared and revered, endorsed and debunked. Some of the problems with establishing globalization's empirical robustness arise from the fluidity of its definition. Peter Urmetzer (2005), for example, argues that a significant problem with the concept is its ambiguity, evident in a "globalization of everything" approach that unjustifiably conflates different processes and leads to unsupported assertions. Nonetheless, he finds three elements common to most definitions of globalization: it involves increased cross-border movement (economic, political, cultural, etc.); it is universally seen to have accelerated in the period between the 1960s and 1980s; and it is seen to cause a weakening of the nation-state (37). This last point, related to perspectives of a new world of "powerful corporations and feckless states" (23), is the most relevant to concerns about the nation-state's ability to regulate its own agricultural system.

Should states indeed be powerless, we can expect international homogenization of weak biotechnology regulatory regimes in spite of differing national visions around the technology, and in spite of subnational opposition and protest against it. Academic arguments associating processes of globalization with a weakening of national regulations, and in fact with the decline of the nation-state itself, can be readily found. Strange (2003) provides a succinct statement of this "loss of state autonomy" position: "Where states were once the masters of markets, now it is the markets which, on many crucial issues, are the masters over the governments of states" (128).

Most quantitative assertions of the globalization argument have now been put aside in favor of representations of globalization as a qualitative break from the past, the key features of which are technology development, corporate concentration and transnationalization, and supranational trade agreements which codify corporate-friendly trade rules on a global scale and contribute to the power shift from nation-states to private capital.

Technology is a key factor in this power shift because it changes the terrain of wealth from that of territory to that of market share (Strange, 2003). In fact, changes in technology have underwritten each of the successive periods of capitalism (Teeple, 2000:175). Society is currently in the early stages of a new technology revolution, including information technologies and genetic modification (Castells, 2000). The significance of the latter for the international restructuring of agriculture and food systems is becoming increasingly apparent. A significant promoter of technological advance is, of course, capital; and capital concentration is another key factor in the shifting state-market balance of power. While capital has operated on an international basis in the past, the transnational corporation [TNC] is argued to play a historically distinct role in the current form of internationalization of capital: "The main actors become the TNCs and all the circuits of capital become global in nature with a distinct global framework; and the national economy—and its associated borders, policies, and programs—becomes a fetter" (Teeple, 2000:179).

In this scenario, attempts to regulate industries—such as the biotechnology industry—run counter to the business-friendly environment necessary to attract capital. Teeple terms this global economic organization a "new reality," a key factor of which is the rise of supranational agreements institutionalizing corporate-friendly free-trade regimes that constrain the regulatory autonomy of nation-states. Globalizationists argue that such agreements provide an enabling framework for the creation of a "single, unified, global market" (Teeple, 2000:179) which has no tolerance for the social priorities of national governments or their citizens.

Other scholars, such as McBride and Shields (1997), contrast this view of crippled states, arguing that any pressures to downsize are domestically motivated. Urmetzer (2005) similarly concludes that there is no case for globalization as an inevitable external force weakening the autonomy of nation-states, but that the concept revisits an age-old debate about state intervention versus laissez-faire capitalism. Most convincing are attempts that blend these polarized perspectives with suggestions that while new constraints on governments do exist, they are relative and not absolute, allowing states a variety of adaptation strategies (see Ó Riain, 2000; Weiss, 1997). For example, while it probably is the case that globalizationists overestimate the constraining nature of supranational agreements (given ample evidence of countries ignoring such constraints), it is also the case that many countries do comply with them, often in ways that are restrictive of, if not directly contrary to, national goals and objectives. To dismiss this reality as based on a case of national false consciousness or weak national will is theoretically unsatisfying.

The American-based social structures of accumulation theory and the French-based regulation theory provide broader insight into this debate through their postulations about long waves of capital accumulation. They propose that capital requires a relatively stable environment for investment—made up of economic and non-economic factors—which is ultimately manifested in the institutions of society. The state plays a central role in determining the nature of such institutions, for example by mediating capital accumulation in areas such as investment in raw materials and organization of the labor process (Gordon et al., 1994). While the two schools differ on what triggers crisis in the system, they agree that restructuring is required to overcome the economic instability created when the institutional structures of a given regime no longer support capital accumulation. This restructuring emerges out of a "complex economic, political and ideological process" (Kotz, 1994:58). The end result is not predetermined, but is likely "shaped by the relative power and the respective objectives of capitalists, workers, and other economic groups" (Gordon et al., 1994:19).

Thus globalization is not seen as an either/or proposition, but as the culmination of active struggle. Since the collapse of the post–World War II period of stability, the "free market agenda of privatization, liberalization, and deregulation has been aggressively pushed on the rest of the world" (Wolfson, 2003:259). Citizen groups and various social movements have been actively seeking to socialize this agenda with respect to globalization in general, and to agriculture and food in particular, as we will soon see: they are contested paradigms. Before pursuing this argument further, I will relate the issues more specifically to agriculture and food.

Food production has much salience for globalization questions, as it is both practically and symbolically tied to national autonomy, an association made explicit by the food regime perspective pioneered by Harriet Friedmann and developed further in work by Friedmann and Philip McMichael (Friedmann, 1992, 2000, 2004; Friedmann and McMichael, 1989; McMichael, 1991, 1992). Rooted in regulation theory's perspective of capital accumulation occurring in distinct regimes, each with particular traits and institutional structures, a food regime is a historically bounded period of norms and expectations that govern all actors in the production and consumption of food on a world scale (Friedmann, 2004:125). It is historical and geopolitical. More specifically, it is an "international political-economic relation linking food production and consumption to dominant historical forms of capital accumulation" (McMichael, 1991:74). Consequently, the shape of the food regime is entirely fixed to globalization trends more broadly.

The following section will briefly outline the key features of the food regime perspective. It will then discuss how potentially mediating activities that occur within nations—and which receive little attention from the perspective—might affect national goals and, consequently, national integration into the food regime.

From the "Settler-Colonial" Regime
to the "Surplus" Regime . . . and Back?

In their seminal paper on the food regime perspective, Friedmann and McMichael (1989) conceptualize two distinct food regimes in the period between 1870 and 1973, with the earlier regime setting the preconditions for the later one. The first, the "settler-colonial" regime, covered the period between 1870 and 1914—a period of British global supremacy. As in the succeeding regime, this regime contained opposing movements between the state system and an international division of labor. On the one hand, colonialism "re-divided the world economy into vertical power blocs" and subordinated the agricultural hinterland to the industrial metropole (ibid.:98). On the other hand, the staples relationship between metropole and settler states facilitated the emergence of the state system.

Long distance trade had existed since 1500, but its impact was mediated by perceptions that governments were responsible for protecting the food supply. The 1846 repeal of the Corn Laws in Britain (import protection for domestic producers, which stabilized prices) effectively ended this role, and "created a trans-oceanic market in basic foods (and an ideology of free trade to justify it)" (Friedmann, 2004:126). In this international division of labor of commercially specialized agriculture, settler states provided wheat and meat for the new proletariat in the metropole, while the resulting staples relationship facilitated these states' national development. At the close of this regime, three agriculture-industry relationships had been established: (1) complementary production (e.g., tropical products from colonies) was replaced by competitive production based on comparative advantage; (2) agriculture began to develop as a capitalist economic sector (through appropriation and substitution); and (3) internationally organized, commercial agriculture remained contained by nationally organized economies (Friedmann and McMichael, 1989:102). These relationships became the predecessors of the agrifood complexes of the second regime.

This first food regime collapsed through the instability of the depression and world wars (Friedmann, 2004). The second arose out of the new economic relationships forged in the post–World War II period. This "surplus regime" saw American-style policy replicated internationally. It was based on "intensive," rather than extensive, accumulation strategies, predominantly characterized by the growing agrifood complexes. This regime also oversaw the completion of the state system through decolonization, balanced against transnational restructuring of agricultural sectors by agrifood capitals, perhaps at the expense of the nation-state system itself. This last point resonates the most with globalization literatures.

Friedmann (1993) proposes that tension between two processes—replication and integration—formed the basis of the surplus regime. Replication occurred through the export of U.S.-style agriculture. After the Second World War, the United States arose as an economic world power. While it was strongly in favor of free trade, it was also motivated to protect its farmers, who were a powerful constituency and who remained dependent on the system of exports established in the earlier regime. Consequently, the United States set a policy agenda that suited these domestic needs—namely through trade restrictions and farm supports—and impelled international trade rules to match. Under American hegemony, this model of national regulation was replicated by other states. Thus agriculture became an exception to international trade rules, creating a "pattern of intensely national regulation" (Friedman, 1993:32).

Countering these nationally regulated agricultures was the increasing integration brought by liberalized trade. Agrifood capital investment "tended to integrate the agro-food sectors of Europe and the United States in an Atlantic agro-food economy" (Friedmann, 1995b: 514). This integration was facilitated by technological advancements such as preservation technologies, which increased the potential for mass production and dissemination. The resulting integrated and internationalized food relations facilitated the rise of the wheat, livestock, and durable foods complexes (Friedmann, 1992). Each complex represented a web of intersecting production and consumption relations.

The most politically significant of the food complexes was the wheat complex. The American agriculture model of industrialization and farm supports made grain surpluses a substantial problem in the United States. This was ultimately resolved through the device of food aid: Public Law 480 allowed the United States to take care of its surpluses by flooding developing countries with heavily subsidized grain. Cheap wheat imports shifted diets and turned these countries from being food self-sufficient to food dependent. Even with the Green Revolution and the export of the high-productivity American model of agriculture, food dependence under shifting world prices increased developing country debt and subsequently increased the pressure to produce agriculture for export to finance it.

The new technologies of the durable foods complex turned food into long-lasting, high-value, manufactured products, and agricultural goods became an intermediate ingredient to these commodities rather than a direct consumer product. Appropriationism and substitutionism (Goodman et al., 1987) were key to this complex. These processes raise even greater control issues than evidenced by the monopolization of markets in the wheat complex (Friedmann, 1993:374).

Regionally unbounded integration is most evident in the livestock complex, which links monoculture crop production with meat production through the capital-intensive feedstuffs industry. While initially nationally organized, "once crop and livestock producers were linked by corporations, inputs in principle could come from anywhere," and quickly did (ibid.:376). The exemplar of the livestock complex is the "world steer," characterized in Sanderson's (1986) investigation of Latin American integration into the cattle and meat sector. Not only does corporate integration link globally sourced feed, production, and consumption, but it also represents the internationalization of industry norms such as preferred cuts of meat and production methods (e.g., confinement feeding), leading to a truly integrated complex. The world steer thus demonstrates a qualitative shift in integration: "international economic integration of the nineteenth century, which relied primarily on commodity circulation, has been supplanted by a holistic integration of the cattle sector in production" (Sanderson, 1986:124).

Geographic inequality has followed hard on the heels of this integration. Latin American countries, for example, forced to exploit their comparative advantage in cattle production, supplant subsistence producers and create grain deficits to produce meat for high-income developed world markets, with significant negative impacts on the domestic population. This inequality is not limited to the livestock complex, however. Production for export of other nontraditional agricultural products—for example, exotic foods and flowers—has similar negative effects (Friedmann, 1993:50). The export of equipment and chemicals for Green Revolution–inspired intensive agriculture similarly benefited major American companies, while creating environmental and social problems in the receiving developing countries (see, e.g., Otero and Pechlaner, 2008).

The food regime concept is thus historical, political, and geographical. Historically, it outlines the trajectories that shape the characteristics of successive regimes. Politically, it is based on differential power relations that divide the world into agrifood power blocs, most notably, northern power and southern dependency. Geographically it outlines an international division of labor in agriculture and food that is reflective of these power differentials. The geopolitical organization of each preceding regime provides the basis for its successor. Following regime collapse, there is a period of crisis which provides openings for influence on the emerging regime (Friedmann, 2005).

The Neoliberal Food Regime

The postwar food regime ended in the 1970s as a result of increasing instability in world markets. Most specifically, a food crisis, triggered by the first grain sales to the Soviet Union in 1972–73, undermined the implicit rules governing surplus. Further instability was caused by the rise in oil prices, the debt crisis, rising export competition, and the collapse of the Bretton Woods system (McMichael, 1992:353). A number of features from the surplus regime were likely to feature prominently in the evolving third regime. The surplus regime's legacy of transnational organization and agrifood complexes appears to have set the course for further agro-industrialization, further "political elimination of barriers to capital" (McMichael, 2004:4), even more flexible global sourcing (and forum shopping), and exacerbated southern dependency. One fundamental difference is that the national regulation and organization of agricultures has become increasingly unviable under the dual pressures of transnational agrifood capital and the rise of supranational agreements.

> Accumulation by agro-food capitals has in the late twentieth century so subdivided and restructured agriculture everywhere—on the basis of highly protective state policies—that the capacities of states and the state system for further regulation are in question. (Friedmann and McMichael, 1989:94)

The emerging third regime essentially specializes many globalization tendencies to food, through globally sourced production and consumption, transnational economic organization, and limited (supranationally organized) regulation. Thus the food regime perspective presents compelling insights into globally directed corporate agricultural strategy, raising the prospect that the high degree of corporate vertical integration will have the potential to render farmers, consumers, and even nation-states increasingly irrelevant to agricultural production decisions. Indeed, McMichael calls the third regime a "corporate food regime" (McMichael, 2005); this term, however, lacks some distinction given that corporations preceded this regime, and "neoliberal food regime" is used here instead. This term specifies that "corporations, like neoregulation, both operate under the impetus of the ideology of neoliberal globalism, which can change with a different configuration of power relations in society" (Pechlaner and Otero, 2010:182).

The rapid introduction of corporate-controlled biotechnologies into agriculture only exacerbates concerns over the antisocial tendencies of the neoliberal food regime. As biotechnology helps deconstruct farm products into interchangeable components, it facilitates substitutionism and global sourcing strategies, while the technological packages offered by life sciences corporations further the integration of input sellers and output purchasers. It is not difficult to make projections of an agricultural future typified by agrifood corporations dictating the production of crops specifically engineered for particular processing sectors. McCain's requirement of genetically standard potatoes for its frozen chips production, for example, "reorganized traditional agricultural communities in Eastern Canada," as monopoly contracts "specifying most aspects of production subordinated family farms and created a monoculture region" (Friedmann, 1992:374). Genetic engineering is a natural extension of this process, where noncompliance simply results in the food processor changing its supply source. Ultimately, if allowed to develop as projected, biotechnology appears likely to underpin the agricultural system of the third regime.

There are also indications that the regional blocs of the second regime are shifting. Southern subordination will likely deepen, as "local farming and informal provisioning" must continue to be converted for the benefit of production for the supply chain (McMichael 2004). This involves, for example, the production of high-value agricultural goods (such as fresh fruit and vegetables) for rich consumers in developed countries (see, e.g., Nagatada, 2006). New agricultural countries are emerging on this production-for-export scene. Perhaps most importantly, new dynamics are also created by agricultural biotechnologies which will affect geopolitical organization. New countries, such as China, are gaining as major producers of GM crops, while others are struggling over the adoption question. At the same time, power brokers, such as the United States and the EU, are setting the international political scene for biotechnology dissemination (as we will see in Chapter 2).

A key motivator of the new geopolitical blocs will depend on the strength of intellectual property rights regimes and on the distribution of ownership of genetic material. Most of the world's germplasm is found in the South, in developing countries, whereas most biotechnology companies are based in northern, developed countries. Intellectual property rights protect the rights of technology developers, not of those who provide the source of the original germplasm, leading some to claim that northern patenting of southern germplasm constitutes a form of recolonization (Shiva, 2001). Others propose that the patentability of life is producing a global hierarchy that reinforces the inequalities between the advanced industrial countries and the less developed countries by favoring those that have gene technology over those that do not (McNalley and Wheale, 1998). For example, "in 1996, the US earned $30 billion from royalties and licenses," while "the South spent $18 billion for buying patented technology in 1995" (Shiva, 2001:28). Monopoly of the world's genetic resources may ultimately be a significant form of global power in a new regime of accumulation.

While the unfolding neoliberal food regime suggests some serious antisocial tendencies, it is not immutable. In later works, both Friedmann and McMichael have played with articulating different variables that could affect it. They have both noted the importance of social movements to neoliberal regulatory restructuring. McMichael (2005), for example, suggests that competing visions of food sovereignty (such as that espoused by La Via Campesina) which attempt to protect small-holder agriculture and those with limited market access to food can act to challenge the WTO-style neoregulatory transformation of a world agriculture (see also Desmarais, 2007). Somewhat less optimistically, Friedmann (2005) suggests that "green" pressures brought by consumers and social movement actors could create a two-tiered food regime: one of fresh, relatively unprocessed, socially sustainable, privately standardized food for rich consumers, and the other producing publicly overseen (by limited regulation), highly processed, and "denatured" edible commodities as food for poor consumers.

McMichael and Friedman provide two obviously contrasting resolutions to social movement activity, yet they both indicate that while agrifood complexes can circumvent national policies, local initiatives are still working to reconnect (or relocalize) production and consumption relations. It is here where I propose the local meets the global: as national law and regulation become forums for local opposition to the technology, the potential of affecting this regulatory framework, and the neoliberal food regime itself, may find its voice.

The Local Meets the Global: A Framework for Analysis?

While the food regime provides a strong framework for conceptualizing the historical relationship of agriculture to the global economy, the conceptualization is accused of operating from a macrolevel pedestal, casting sweeping generalizations about the internationalization of agrifood production. Le Heron and Roche (1995), for example, propose that the food regime conceptualization captures the "interplay between evolving accumulation and regulation processes," but is "surprisingly silent on geography, accenting historical over geographic insight, except in coarse geopolitical terms" (24). Essentially, it is accused of lacking an explicit treatment of the national and local regulatory dimensions that affect differential integration into the food regime.

A number of case studies can be found that highlight social factors that affect regulatory and legislative processes, and that consequently compromise integration into the food regime, or globalization project more generally, however. Le Heron and Roche (1995), for example, identify the interplay of globalization and sustainability ideas evident in local regulation and political negotiations as key to New Zealand's differential incorporation into the third food regime. Moran et al. (1996) draw on New Zealand and France to argue that it is possible for nations to retain characteristics distinct from global agrifood transformation patterns. In their examples, farmer strategies (such as producer cooperatives) and their effect on commodity chains have been able to influence differential national integration into the regime.

Numerous other studies document differentiation from the key tendencies identified by the globalization of agriculture: global sourcing, deregulation, and corporate restructuring. This differentiation is often the result of some form of local resistance. Constance et al.'s (2003) investigation of global sourcing in chicken production in Texas reveals that the success of corporate strategies is not predetermined; rather, the decoupling of community benefits from economic development can spark significant resistance. Novek's (2003) investigation of hog farming in Manitoba found the globalization logic disrupted by conflict and grassroots opposition. Wells' (1997) study of industrial restructuring in California's strawberry industry found a counterintuitive rise and fall of sharecropping to globalization trends, suggesting that restructuring may be an "intentional tactic" by some producers to "help them mitigate locally experienced political challenges" (250), rather than responses to the world market.

Community responses to globalization in agrifood are not the only forces shaping the face of globalized agriculture. Conducting a case study of four significant agrifood commodity chains, Friedland (2004) concludes that "globalization—in its spatial sense—is extremely uneven, possibly especially in agrifoods" (14). In short, there are mixed tendencies to agrifood concentration and globalization, depending on a raft of social and technical factors, from the expansion of refrigeration capacity to the distribution of income. True to appropriation and substitution conceptualizations, capital expands where it can, when it can, and how it can: but it can't always.

Goodman and Watts similarly find that there is significant international evidence of different patterns of replication and integration. They ask:

> Has the shift from post-war stability to the crisis of the post-1970 period been too readily characterised as a historic victory of transnational capital without sufficient attention to massive instabilities (productive and institutional) and frictions (active and passive resistance) within the "new" internationalization of agriculture? (1994:21)

In essence, they critique the food regime perspective for overemphasizing the homogeneity of the spatial reorganization of agricultural production. Complementing scholars such as Urmetzer (2005), Goodman and Watts argue that supranational agreements are unlikely to subordinate the nation-state. They find significant state contravention of trade agreements and provide evidence from the North American Free Trade Agreement to argue that deregulation in one sphere produces re-regulation in another, revealing "how unlikely, in fact, is a free trade regime" (23).

In sum, these cases provide no shortage of exceptions to the global agrifood tendencies outlined by the food regime perspective. If the revelations of these cases were to be expressed in one sentence, it would be this: "Social forces count." While not highlighted, such ideas about national difference and consequent differential integration into the food regime are actually implicit in the perspective. Friedman, for example, notes that in the postwar restructuring, "states replicated the US regulation of national sectors, but adapted policies to their locations in the food regime" (1993:32). This awareness is even more explicit in a recent resurgence of scholarship on the food regime perspective (see, e.g., volume 26 of *Agriculture and Human Values*).

It is important not to swing the conceptual pendulum to the other extreme, however, forgoing the identification of broad patterns for the sake of conceptual specificity. Goodman and Watts, for example, discount the perspective's ability to specify the mechanisms leading to a stable third regime: they argue that it paints a highly flawed core-periphery dependency model which hypothesizes crude North-South agrarian restructuring and overlooks southern differentiation. While such differentiation certainly exists, overly emphasizing it risks neglecting commonalities that have serious implications, particularly for the more vulnerable. If Sanderson's (1986) case study of cattle production in Latin America can be generalized in any way, for example, and indications are that it can, the repercussions of such changes should not be overlooked:

> The emergence of the "world steer" has shifted power away from the primary producer; it has "disarticulated" consumption from the national economy (and certainly from the rural economy); and, it has created negative effects in foodstuff production and land tenure. The combined effect has been to help make the poor poorer, the malnourished more malnourished, and the heralded era of "rural development" a bureaucratic nightmare. (146)

The appearance of these effects with any sort of regularity is sufficiently troubling that regional differentiation hardly seems justification for neglect.

At heart, such concerns with national differentiation versus a global food regime are really about how theory can accommodate the meeting of the global and the local. One resolution may be provided by Lourdes Gouveia's (1997) case study of globalization restructuring in Venezuela. Gouveia concludes that while efforts to enroll Venezuela in the globalization project have been heterogenous and at times ineffective, "the fingerprints of the neo-liberal agenda can still be detected" (316–317), often with significant impacts on the population. While Gouveia's study is significant in its own right, perhaps its most valuable contribution is her well-balanced perspective on the differentiation problem. Pointedly, she questions whether the eagerness to document heterogeneity might become an end in itself, leading social scientists to fail to identify relatively stable institutional arrangements that profoundly affect people's lives. She suggests that while we should be "mindful" of deductivist characterizations of globalization, this should not prevent us from assessing the very important transformations occurring in the relationship between the state, the market, and civil society:

> The analytical task does not come to an end with the discovery of diversity, heterogeneity, or the fact that all actors have some degree of power. . . . [I]t is important to complete the analytical loop and return to the macro level for a simultaneous interrogation of data and historical constructs to determine whether, despite diversity, broader sociostructural changes can still be identified. (316)

In practical terms, striking the macro-micro balance would require an empirical focus on local interactions while remaining attentive to how these local interactions fit into broader patterns. Negotiating this balance appears to have advanced much further in globalization literatures other than those specific to food.

McMichael (1996), for example, argues that globalization is a political project, and we are best served by problematizing it as a new set of institutional and ideological relations whose application is partial and contested: "Communities scramble to reposition themselves either through finding niches in a new global economy or through resistance to global pressures" (25). Séan Ó Riain (2000) similarly argues that globalization is a political project and that while states "may be threatened," they are also "the primary actors that will continue to shape the process itself" (206). Echoing food regime scholars, Ó Riain posits that the particular state-market-society relations institutionalized in advanced capitalist countries post–World War II (which affected relations far beyond them), and that weakened under globalization, now must be forged anew. He argues that not all states will homogeneously arrive at the same set of relationships, however: "The relations among them are inevitably tense, due to the inherent dilemmas of reconciling market, society, and state in a capitalist economy. . . . The way in which these three spheres shape one another be-comes the central determinant of an economy's fate under globalization" (191).

While each state's response will depend on national differences, this does not mean it is completely unpredictable how they will be integrated into the new regime, and that we should simply revert to documenting heterogeneity, as Gouveia warns against. Ó Riain, for example, considers four dominant models of state-market interaction—liberal states, social rights states, developmental states, and socialist states—to provide a semi-structured perspective of global restructuring without reverting to a one-dimensional North-South new regime scenario. Echoing McMichael, he argues that globalization "is a politically contested process in which dif-ferent state-market models of interaction come into conflict locally, na-tionally, and transnationally" (ibid.:188). Consequently, how different states navigate the conflict will ultimately affect the globalization project itself: "To the degree that nation-states resist the specific forms of inte-gration being promulgated through the World Trade Organization, na-tional regulatory apparatuses could impede accelerated global economic concentration" (Friedland, 2004:15).

In the same manner that the globalization project is not globally ho-mogenous, however, neither are the intentions of any given nation-state. Rather, these "intentions" themselves (whether toward a globalization ideology or specific agrifood policies) are ultimately achieved through many varied struggles. Gouveia builds on McMichael's perspective of globalization as partial and contested to suggest that a multitude of actors are involved in reformulating the relationship between market, state, and society, and consequently multiple globalization projects "intersect, modify, or contradict one another" (1997:309). The result is a globaliza-tion that arises out of contradictory results: some promote the ideologi-cal underpinnings of the neoliberal thrust of globalization, and some are counterthrusts that undermine them. The biotechnology-focused neolib-eral food regime is one such project. What remains to be seen is whether and how these "thrusts" will shape national priorities around it. Given the importance of state-supported intellectual property rights and laissez-faire regulation for the development of commercial biotechnologies (as we shall see in Chapter 2), factors which affect these will likely be vital to its future development.

Factors for Resistance: Corporate
Biotechnology as Contested Project

The contours of the neoliberal food regime are thus fairly demarcated, but far from inevitable. Given the high contestation around biotechnology, its position is potentially even more vulnerable, with opposition to the technology readily apparent both nationally and internationally. Similar to Polanyi's (2001 [1944]) concept of the double movement, if the market drive to expand agricultural biotechnologies becomes socially irrational, there will be a consequent societal protective movement. McNalley and Wheale, for example, argue that the social reordering caused by modern biotechnology and biotechnological patenting has provoked an opposing social movement which challenges the unspoken transfer of power by eroding the "apparent neutrality and anonymity" that lies behind it (McNalley and Wheale, 1998:326). "Its challenges force the "bio-industrial complex" to produce justifications, for example, for its regulatory policies and corporate strategies, justifications which so often reveal the inequity of its conduct, structure and performance" (325).

Ulrich Beck (1992) uses the term "sub-politics" to describe how groups apply pressure in nonpolitical forums when social futures appear increasingly independent of local democratic input. In the face of weak national regulations, for example, transnational social movements have become an important source of direct challenge to corporate power (Holzer, 2001). Transnational consumer boycotts are one such example. Friedmann (1993) and Busch and Bain (2004) note that wealthy consumers can directly confront companies with their demands, thus "substituting consumer demand for citizen demand, market accountability for governmental accountability" (335). In this manner, national complicity in the neoregulatory refashioning of the food regime notwithstanding, corporate entities can still be "confronted with a situation where the legitimacy of their operations may be challenged no matter what their legal status" (Holzer, 2001:80).

With respect to food, there is ample evidence of all kinds of movements resisting the parameters laid out by the neoliberal food regime and the emerging prominence of biotechnology within it: sustainability movements emphasizing local production and consumption; consumer demand for organics; citizen resistance to intensive livestock production; and subnational initiatives banning GMOs, to name a few. Incidents of contestation, some locally successful, are abundant. Most specifically, the earlier case studies demonstrating national deviation revealed common causes and sites of resistance to the globalization paradigm. Community-level resistance featured prominently, particularly where deregulation left communities exposed to corporate strategies that had negative community impacts. In their chicken production case study, for example, Constance et al. concluded:

> The opening of local communities to globalization lacks institutions capable of buffering the unwanted consequences of the growth of capitalism and controlling its most powerful actors. This situation engenders fierce resistance and creates a contested terrain in which corporate power is resisted and redefined even in conditions in which the demands of local residents are grounded in discourses parallel to those of corporations. (2003:117)

Of particular salience here is that the "parallel discourses" referred to are discourses around property rights. The loss of property value resulting from the high-intensity production was not addressed in the corporate plans, and, ultimately, "it was the issue of property that served as a catalyst for resistance" (116). Similarly, this issue of property is finding significant resonance in agricultural biotechnology-related litigation. Given the proprietary issues raised by GM technologies, both with respect to patent infringement and contamination, such legal issues are key sites for agitation for national regulatory input.

The use of legal mobilization as a social movement tactic is a difficult theoretical area, however. In part, this is because there is a dearth of scholarship that successfully links legal scholarship with social movement scholarship (McCann, 2006). In part, this is because the scholarship that does exist is overly focused on limited causal variables and outcomes. For example, social movement scholars are often analytically attentive to the factors that affect social movement mobilization, or, alternately, on how social movement mobilization affects policy outcomes. The false sense of unidirectionality resulting from such limited causal or outcome-focused analyses fails to accommodate any reciprocity through which changes in the broader society affect legal mobilization and vice versa. Further, these analyses usually fail to address the deeper question of how the mobilization affects the goals of the group—that is, how "law matters" for social movements (ibid.).

It is increasingly clear, however, that the consequences of such legal mobilization extend beyond "their intended or policy effects," and include broader consequences, for example, by altering the cultural environment

(Guigni, 2008:1591; see also Earl, 2004). It is here that movements "can have their deepest and lasting impact" (Guigni, 2008:1591) through spin-off movements, spillover effects between movements, the diffusion of ideologies, and other impacts on further actions for social change (Whittier, 2004:532). These are the often neglected but "radiating" effects of the use of legal action (Galanter, 1983, cited in McCann, 1994:10). Thus the effectiveness of a course of legal action on social change cannot be assessed just by its legal end result, although this is certainly an important factor.

With respect to the legal outcome itself, there is certainly ample evidence in support of the many critical visions that conceive of the law as a tool for the powerful. Consequently, conclusions regarding the legal effectiveness of such social movement actions are often negative (McCann, 2006). Marc Galanter's seminal article (1994 [1974]) carefully articulates how the "haves" come out ahead of "one-shotters" in litigation through a variety of processes that favor the former. These processes will unquestionably affect the outcome of any litigation between biotechnology developers and farmers. At the same time, while certainly biased, the outcome is not entirely predetermined. Bourdieu (1987), for example, conceptualizes the "legal field" as "the specific power relations which is its structure and which order the competitive struggles," and, at the same time, "the internal logic of juridical functioning which constantly constrains the range of possible actions" (816). Power is thus a factor, but not necessarily an insurmountable one.

This conceptualization of the legal field can be extended to provide some insight into the implementation stage as well. Moran et al. (1996), drawing on Clark (1992), for example, suggest that regulation is a social practice that occurs in an economic, cultural, and geographic context, and evolves from overlapping contests for power. Significantly, they emphasize the role that citizen interaction with the law plays in such regulation:

> Existing policies and practices are actively negotiated and renegotiated
> (respectively) by a range of lobby groups. Social actors, whether in-
> volved or not in the first round of formulation, create the pressure to re-
> formulate or repeal legislation if they do not like the original legislation
> in the courts, planning processes and its social application. (249)

This is the difference between the black-letter "law on the books" and the "law in action" (Greenbaum and Wellington, 2002; Sumner, 1979). Law and society are mutually influencing, and "changes in society's values and public opinion can feed back into the legal system and affect the prospects for law reform and enhance the effective implementation of legislation" (Coglianese, 2001:86). This emphasizes the importance of taking a broader analytical scope than a strict reading of legal change could provide.

At this juncture, biotechnology litigation appears unlikely to outright reverse the national pro-biotechnology development direction, although it remains possible. Even more likely, however, is that such litigation will impact its still-contested regulatory framework or heighten social opposition to the issue. The potential for institutionalization that such legal resistance brings is another key for the extent of social impact. As Friedmann and McMichael note: "Ultimately the success of local projects depends on their combination and co-ordination at higher levels, to replace the policies (and confront the powerful interests associated with them) favouring a global orientation of both production and consumption" (1989:113).

National biotechnology-related policies can be affected by higher court rulings on the patentability of life, ownership of self-reproducing inventions, infringement, liability over the involuntary presence of patented material, and monopoly and antitrust issues. Such rulings could have a major impact on whether a location is more or less favorable for the biotechnology industry, and on the strategies it has available for economic development. Local-level resistance to loss of control through expropriationism in the legal forum are thus strong possibilities for opposition that could have broader effects, potentially even impacting the unfolding global food regime.

Conclusion

A perspective of the globalization of agrifood that is partial and contested, rather than inevitable—while falling into roughly predictable geopolitical patterning—is emerging. This study thus proceeds on the theoretical assumption that there is globalization, but it is differentiated globalization, and this applies to the globalization of food and agriculture no less. The neoliberal food regime is one manifestation of globalization, and corporate agricultural biotechnologies are poised to become important features of its fullest expression. The conceptualization of globalization as a project, or projects, provides a means of considering how biotechnologies, as one such project, are subject to opposition and contestation over the terms of their integration. While the food regime is ultimately a global phenomenon, the negotiation of its terms occurs at the national and subnational levels, and is influenced by a number of local factors.

Biotechnology's introduction into agriculture has already initiated a number of local-level changes, ones which I suggest Goodman, Sorj, and Wilkinson's (1987) conceptualization of capital's piecemeal integration into agriculture seems unable to capture. Rather, agricultural biotechnologies' proprietary aspects suggest that a new capital accumulation strategy premised on legal means of accumulation may be occurring, one for which I propose the term "expropriationism." Property issues are an important factor in motivating local opposition, however. The expropriationism associated with the introduction of agricultural biotechnologies may be a similar motivating factor. While local differentiation is important in itself, the goal is to "complete the analytical loop," as suggested by Gouveia. If it is indeed possible for the institutional manifestation of local resistance projects to redirect or even replace the policies of globalization—whether through lobbying, social movement pressure, or the more institutionally direct path of litigation—then there is the potential for such activity to affect the global food regime.

CHAPTER 2

The Coming of the Third Regime?
Agricultural Biotechnology Regulation
in Canada and the United States

Introduction

As noted in the previous chapter, Canada and the United States are the only industrialized countries with significant GM crop area (51% of global area). Outside of these, the majority of biotechnology production area is in developing countries (48% of global area), most notably Brazil, Argentina, and India. The European Union [EU], by contrast, has minimal biotechnology. The most notable production in the EU occurs in Spain (0.07% of global area) (percentages calculated from James, 2010). The rate of adoption in developing countries is rapidly surpassing that of developed countries, however. Between 2009 and 2010, the rate of growth in developing countries was 17% versus 5% in industrialized countries (ibid.). Nonetheless, the industrialized countries are still the drivers of important regulatory paradigms which set the tone for technology development and diffusion. Thus their regulatory positioning (and counterpositioning) has great significance for the shaping of the global food regime.

The physical risks of genetic modification have been documented in many forums. These include risks to human health or the environment, such as risks related to allergenicity and toxicity (e.g., introducing new proteins into the human diet, some with pesticidal properties), unexpected repercussions from crop modifications (e.g., insecticidal crops having negative impacts on nontarget species), and the risk of mutations of GM crops (e.g., if they cross with similar but undesirable species, potentially creating superweeds). Such risks are deeply tied to concerns over the unprecedented intervention in life and to concerns that scientific uncertainties preclude a full understanding of the potential repercussions of this intervention. The fact that any intervention in life creates self-reproducing products vastly exacerbates concerns over such risks. Lastly, while the first generation of biotechnology traits focused on agricultural production benefits (e.g., herbicide tolerance and insect resistance), the second generation emphasizes applications designed to benefit consumers, such as nutritional enhancement or medical benefits. Of particular concern, these second-generation traits include pharma-crops—crops engineered to have pharmaceutical properties, such as blood-thinning agents or contraceptives.

The point is not to emphasize the risks of biotechnologies over their potential benefits, but to outline areas that need regulation. Given the range of risks, limitations in knowledge, and diversity of applications, the rapidity of adoption exacerbates the need for regulation, as does the sheer volume of the technology in the environment. Of course, there are also the additional social aspects of the technology, which have received far less attention. On the one hand, a strong proprietary framework is necessary for the development of the industry. On the other hand, providing intellectual property protection on self-reproducing "inventions" leads to inevitable clashes between farmers and technology developers. Once again, legislation addressing these clashes would seem paramount to maintaining the social benefits of the technology.

Just at the time when the need for national regulation would appear to be at its most pressing, however, critics charge that international trade agreements preclude autonomous action. Is this really the case? The first section of this chapter outlines international regulation relevant to agricultural biotechnology. The second looks at biotechnology regulation in the United States, Canada, and the European Union. Rather than being helpless pawns, the former two, in tandem, are creating a low-regulation, pro-biotechnology bloc, while the latter demonstrates significant independence from these trends. Given the global nature of agriculture production and marketing, it is necessary to consider the impact of the EU's antithetical regulatory approach on the shaping of biotechnologies' role in the neoliberal food regime. The third section comments on these trends and outlines important domestic factors—both inside and outside the legal forum—that can potentially influence national priorities around agricultural biotechnologies.

International Regulation of Biotechnology

The General Agreement on Tariffs and Trade [GATT] was drafted in Geneva in 1947 and came into effect in 1948. Since that time, international trade regulation has evolved through numerous rounds of negotiations involving increasingly more member countries. The Uruguay round was the eighth round, and through it the World Trade Organization [WTO] was brought into existence in 1995. Agriculture was a critical component of the Uruguay round, as the United States and its allies promoted agricultural trade liberalization in an attempt to curry southern buy-in (through promised expansion of access to northern markets), prevent loan default, and gain important concessions in other forms of trade liberalization and intellectual property protection (Buttel, 2003:155). Thus the WTO is pivotal to the internationalization of agricultural biotechnologies, both for its concern with agriculture and for its concern with intellectual property protection.

As an international trade organization, the WTO fosters trade among nations. Toward this goal, the WTO encourages member states to make their national regulatory standards conform to international ones. This push to regulatory harmonization is the basis of much concern regarding the demise of national regulatory autonomy. Three WTO agreements highly relevant to the regulation of biotechnology concern the conditions under which a country can set food safety policies (the Agreement on the Application of Sanitary and Phytosanitary Measures [SPS Agreement]); technical regulations and industrial standards (the Agreement on Technical Barriers to Trade [TBT Agreement]), and intellectual property protection (the Agreement on Trade-Related Aspects of Intellectual Property Rights [TRIPS Agreement]). Canada, the United States, and the European Union are all members of the WTO and thus subject to its agreements.

The purpose of the SPS and TBT Agreements is to address concerns that national policies ostensibly adopted for the purpose of protecting consumers could act as discriminatory trade barriers. These agreements set out guidelines to prevent this. The SPS Agreement, for example, is related to food safety and animal and plant health. Article 2.2 of the SPS Agreement states:

Members shall ensure that any sanitary and phytosanitary measure is applied only to the extent necessary to protect human, animal or plant life or health, is based on scientific principles and is not maintained without sufficient scientific evidence. (Cited on Codex Alimentarius website; hereafter Codex)

Further, the agreement states:

> To harmonize sanitary and phytosanitary measures on as wide a basis as possible, Members shall base their sanitary and phytosanitary measures on international standards, guidelines or recommendations, where they exist, except as otherwise provided for in this Agreement. (ibid.: Art. 3.1)

While countries may set their own regulations, they must be "scientific" and can only apply regulatory measures to the extent necessary for protection. Consequently, they are encouraged to use international agreements that aid in establishing benchmarks for what is scientifically necessary. The Codex Alimentarius, an international standards–setting agency presided over by the Food and Agriculture Organization [FAO] and the World Health Organization, is one such source of benchmarks. The agency coordinates the design and promotion of food standards, guidelines, and codes of practice for the dual purpose of protecting consumers and harmonizing standards to ensure fair trade practices and facilitate international agricultural trade (Codex). The SPS Agreement accepts Codex standards as "scientifically justified" (ibid.); thus, adherence to them can prevent trade disputes.

Codex standards are also used in regional free-trade agreements such as the North American Free Trade Agreement [NAFTA], where they set the basic requirements to be met by member countries. Nonetheless, even the most well-intentioned "scientific" basis of decision making is subjective, and disagreements on the making of these standards occur. For example, an important biotechnology-related issue long under consideration at the Codex is whether foods containing GM organisms should be labeled for consumers. This is a hot issue for marketing GM products, given that without such labels there is no way for consumers to boycott or avoid them once they have been approved. Consequently, the United States is strongly against such labeling. The European Union, on the other hand, is strongly in favor of it, and has already adopted labeling legislation, the Codex notwithstanding.

While the SPS and the TBT Agreements are relevant to agricultural biotechnology because they ensure that GM products have access to markets, the TRIPS Agreement is designed to ensure the economic returns that make the development of such products possible in the first place.

Biotechnology developments are high risk, require large capital investment, and are very slow to reach commercialization. Thus, the patent portfolio itself often becomes the means of garnering investment and staying in business (Dutfield, 2003:153). Consequently, intellectual property protection is at the heart of the biotechnology industry.

The WTO's TRIPS Agreement requires member countries to have a system in place for the protection of intellectual property, with some implementation deadline variation between developed (1 year), developing (5–10 years, depending on initial protections), and least developed (11 years) countries (WTO, "Legal Texts"). The agreement designates that patents must be available "for any inventions, whether products or processes, in all fields of technology, provided that they are new, involve an inventive step and are capable of industrial application" (WTO, "Trade Related Aspects": Art. 27.1). However, there are some allowable exceptions, such as for the protection of "*ordre public* or morality" (Art. 27.2). Most notable here, however, is Article 27.3b, whereby "Members may also exclude from patentability"

> (b) plants and animals other than micro-organisms, and essentially biological processes for the production of plants or animals other than non-biological and microbiological processes. However, Members shall provide for the protection of plant varieties either by patents or by an effective *sui generis* system or by any combination thereof.

Consequently, on the basis of Article 27.3b, plants are an allowable exemption from patent protection. Unfortunately, this section has also been the subject of much controversy and debate, not the least because a definition of what would constitute an "effective sui generis system" was never provided. One sui generis system acceptable to the TRIPS (in the same manner that Codex standards are acceptable to the SPS) is provided by the Convention of the International Union for the Protection of New Varieties of Plants [UPOV]. The UPOV model for plant breeders' rights is enacted in a number of countries, such as the United States and Canada.

The first UPOV convention, in 1961, provided a means to grant intellectual property protection over plants while allowing for broad exceptions for farmers and breeders, which would not be available under a patent system. Subsequent revisions have tightened the convention. The majority of signatory countries currently belong either to UPOV 1978 (as does Canada) or to UPOV 1991 (as does the United States). The later version significantly improves the protection offered to technology developers: it increases the duration of monopoly protection from a minimum of 15 to 20 years; makes previously mandated farmer and plant breeder exemptions a matter of national choice; and removes the prohibition on double protection, whereby any species eligible for plant breeders' rights protection were ineligible for patent protection (Dutfield, 2003:191).

The WTO and its affiliated organizations and agreements provide the most well established and the most influential global regulations affecting agricultural biotechnologies, albeit with limited scope. Concerns with genetic resources related to issues like equity and biodiversity have motivated other international agreements with nonmarket agendas, such as the Convention on Biological Diversity [CBD] and the International Treaty on Plant Genetic Resources for Food and Agriculture [ITPGR], the former of which is already of some import.

The CBD was negotiated under the United Nations Environmental Protocol. Over 150 countries signed it in 1992, and it entered into force in 1993. Canada, the United States, and the European Union are all signatories. The CBD has three main goals: "the conservation of biological diversity, the sustainable use of its components, and the fair and equitable sharing of the benefits from the use of genetic resources" (CBD, "Sustaining"). At the heart of the CBD is a concern with protecting our common heritage of genetic diversity, and it advocates the use of the precautionary principle toward any threats to this diversity: "lack of full scientific certainty should not be used as a reason for postponing measures to avoid or minimize such a threat" (ibid.).

In 2000, the parties to the CBD adopted a supplementary protocol, the Cartagena Protocol on Biosafety, which focuses on protecting biodiversity by managing the risks from living modified organisms. The protocol outlines a number of procedures related to living modified organisms and mandates the establishment of the Biosafety Clearing-House to "facilitate the exchange of scientific, technical, environmental and legal information" and to provide assistance to parties to implement the protocol (CBD, "Background"). The protocol's Advanced Informed Agreement outlines information and decision-making requirements for importing and exporting living modified organisms. Most importantly, Article 10 reaffirms the precautionary principle in import decisions:

> Lack of scientific certainty . . . shall not prevent that Party from taking a decision, as appropriate . . . in order to avoid or minimize such potential adverse effects. (CBD, "Background": Art. 10.6)

Consequently, the CBD takes a markedly different approach from that of the WTO. This is no doubt partly due to the fact that nongovernmental agencies [NGOs] had input into the CBD negotiations, whereas such groups were excluded from the negotiations that led to the creation of the WTO (Buttel, 2003:163). It is thus not surprising that while the United States signed the original CBD, it did not ratify the agreement or sign on to the Cartagena protocol. Canada ratified the CBD in 2001, and while it also signed the protocol that same year, it has not ratified it.

The FAO is generally supportive of biotechnology and of its potential to increase production and alleviate hunger and food insecurity. At the same time, however, it acknowledges that biotechnologies have some environmental and health risks, and that there is an imbalance in benefits given that biotechnologies are produced by the private sector for those with purchasing power in high-income countries (FAO, "FAO Statement"). Consequently, the FAO also oversees the International Treaty on Plant Genetic Resources [ITPGR], which was adopted in 2001 and entered into force in 2004. The ITPGR is concerned with the fair and equitable sharing of benefits from plant genetic resources. It has 110 signatories, including Canada, the United States, and the European Union, although only Canada has ratified the agreement. Similar to the CBD, the ITPGR has as key components conservation and sustainable use. In fact, the ITPGR claims to be "harmonized" with the CBD, although its concerns are more specifically aimed at the food and agriculture industries (Nottenburg and Sharples, "Key Organisations": n.p.), and its approach to its goals appears more weighted to "public funding and dialogue" (FAO, "FAO Statement"). The impact of the ITPGR remains to be seen. Ultimately, the FAO's role is only to provide "assistance" to member countries: the "responsibility for formulating policies towards these technologies rests with the Member Governments themselves" (ibid.). Of course, this "assistance" can be quite significant when guidelines are incorporated into international agreements, as we see with the Codex Alimentarius.

It is beyond the scope of this chapter to further assess the potential of these agreements or the politics behind their creation and implementation. At the present time, the WTO agreements are the most powerful international regulatory driving force for agricultural biotechnologies, to the detriment of nonmarket concerns, many would argue. With this context in mind, I will now turn to the regulation of agricultural biotechnology in the United States, Canada, and the European Union. Given that the United States pioneered the biotechnology industry and remains at the forefront of GM adoption acreage, it can likewise be expected to have had a head start on the regulation of the technology. Thus I will address it first.

National Regulation of Agricultural Biotechnologies

The United States

With respect to its established uses, the United States is at the forefront of a full-scale transgenic transformation of some of its key agricultural crops. For example, in 2008, when transgenics made up 24% of global corn, they made up 80% of U.S. corn; 46% of global cotton versus 86% of U.S. cotton; and 70% of global soybeans (from 64% in 2007) versus 92% of U.S. soybeans (in hectares; global statistics are from James, 2008; U.S. statistics are from United States Department of Agriculture [USDA], Economic Research Service [ERS], 2009). In some U.S. states, adoption rates are even higher: for example, already 95% of the corn and 97% of the soybeans in South Dakota, and 94% of the soybeans and 93% of the cotton in Mississippi, are transgenic (USDA, Economic Research Service [ERS], 2009).

Consequently, the United States is the undisputed leader in biotechnology with respect to its early experience with the technology, its amount of dedicated crop area, its degree of transgenic transformation in a number of key crops, and its relative power in comparison with other biotechnology-producing countries. Further, 75% of publicly traded biotechnology companies are based in the United States, in comparison with 15% in Europe and 8% in Canada (ETC Group, 2005b). Given the dominance of the role that the United States has traditionally played with respect to global agriculture, the manner in which the country regulates the new technology has added salience.

Biotechnology regulation began early in the United States. Agricultural biotechnologies were commercialized in the mid-1990s, but the promise and peril of other GM applications had already entered the scene well before then. Federal regulatory oversight of biotechnology began in 1976 with the National Institutes of Health (NIH) Recombinant DNA Guidelines, however, they were of limited applicability, and voluntary outside of institutions with NIH grants (Marchant, 1988:168). Medical biotechnology produced the first approved GM drug (human insulin) in 1982, but the release of GM organisms into the open environment created additional considerations. Such an application—for the "ice-minus" bacteria to reduce frost damage in strawberries and potatoes—was vetted at about the same time that GM insulin was approved. It sparked citizen opposition, legal challenges, delays, and ultimately a "turf battle" over regulatory jurisdiction (169). "Ice-minus" ultimately became the first legal, environmental release of a GM organism, but this did not occur until 1987 (ETC Group, 2003). Prior to this, in 1984, the U.S. administration had initiated an interagency working group to try to resolve the regulatory squabbles.

The economic potential of the industry had already been recognized. Consequently, the goals of the working group were to ensure health and environmental safety "while maintaining sufficient regulatory flexibility to avoid impeding the growth of the infant industry" (United States, *Federal Register*, Office of Science and Technology Policy [US, FR, OSTP], 1986:23302). The main policy question put before the group was simple: "whether the regulatory framework that pertained to products developed by traditional genetic manipulation techniques was adequate for products obtained with the new techniques" (23302).

The working group concluded that "for the most part, these laws as currently implemented would address regulatory needs adequately" (ibid.:23303). This conclusion was backed by an overall policy position regarding the "substantial equivalence" of GM products to conventionally produced ones. Consequently, biotechnology products were to be regulated, like their conventional counterparts, according to the uses to which they were put (their end-products) rather than the process by which they were produced.

The outcome of the working group was the 1986 "Coordinated Framework for the Regulation of Biotechnology." This framework outlined the roles and legislation by which current agencies would regulate biotechnology. It thus used a "mosaic of existing federal law" (ibid.) and provided detailed information about which aspects of biotechnology were to be regulated by which agencies, and under which statutes. Allowing for departmental shifts and amendments, the regulatory "mosaic" of the "Coordinated Framework" remains the basis for U.S. biotechnology regulation today, with the regulatory responsibility primarily shared by three agencies: the Food and Drug Administration [FDA], the Environmental Protection Agency [EPA], and the United States Department of Agriculture [USDA].

Broadly speaking, the FDA is responsible for food, feed, and food additives, or the safety of consumed products. Its policy "is based on existing food law, and requires that GM foods meet the same rigorous safety standards as is required of all other foods" (USDA, Animal and Plant Health Inspection Service [APHIS] website). The EPA is responsible for pesticides and novel micro-organisms. Thus it ensures the safety of chemicals and biological pesticides through setting tolerance limits; for example, the tolerances for herbicides on herbicide-tolerant crops and for pesticides in insect-resistant plants, two key GM applications. The USDA is the lead agency for agricultural biotechnology. Under the USDA, APHIS is responsible for protecting U.S. agriculture from pests and diseases and for monitoring the introduction of new organisms into the environment. Given the system of regulating according to end use, some products require regulation by more than one agency due to the mixed application of some biotechnology products (e.g., food and pesticide). For example, while the EPA regulates the pesticide tolerance limits of plant-incorporated protectorants in insect-resistant crops such as Bt corn, the actual plant itself (e.g., its importation, transportation, and planting) is regulated by the USDA through APHIS.

Genetically modified products are "regulated articles," and APHIS has jurisdiction over the release of such articles into the environment. Field trials of new GM crops can be authorized under either the permit or the notification system. Permits are required for high-risk crops, such as pharmaceuticals or industrial compounds. Notifications are required for "familiar crops and traits considered to be low risk" (APHIS website). Both systems require the submission of protocols to meet performance standards, and, while field trials are ongoing, crops are subject to monitoring and compliance inspection by APHIS. Following field-testing, application can be made for nonregulated status, and APHIS assesses the application based on studies and data supplied by the applicant. While both regulated and nonregulated products can be commercialized, once deregulated, a product no longer requires any monitoring. The first such deregulation in the United States was Calgene's Flavr Savr tomato in 1992.

The concept of substantial equivalence underlying the U.S. regulatory framework is readily apparent in the policies of each agency. For example, the crops and traits considered "familiar" by APHIS, and thus considered suitable for authorization under the vastly streamlined "notification" process, are the vast majority: in 2004, 97% of GM field trials were notifications (Pew Initiative on Food and Biotechnology [Pew], 2005:2). The FDA has a similar approach. It has authority over substances that are added to food, either as additives (which require pre-market review and approval) or as substances that are "generally recognized as safe" [GRAS]. In general, the FDA considers GM foods to be GRAS: "Many of the food crops currently being developed using biotechnology do not contain substances that are significantly different from those already in the diet and thus do not require pre-market approval" (US, FR, OSTP, 1986:23310).

The determination of a particular product being GRAS is not made by the FDA, however, but by the food manufacturer, although a voluntary "affirmation" process does provide guidance (Pew, 2001:20).

Lastly, in marketing, GM products are again considered the same as their conventionally bred counterparts. The labeling of products as containing GM material is voluntary in the United States, although the few such attempts made—such as for rBST [recombinant bovine somatotropin]-free milk—faced significant anti-labeling litigation (see, e.g., Barboza, 2003). Concerned consumers thus have no means of avoiding GM foods, outside of purchasing organic products.

Canada

Statistical data regarding the adoption of GM crops in Canada are far more difficult to obtain than for the United States, no doubt in part due to the drastically smaller size of the industry. In addition, the Canadian regulatory framework for biotechnology was developed about seven years after the American one. While there are differences, a fair amount of similarity between the two is nonetheless readily apparent.

The Canadian government's official relationship with biotechnologies began in 1983 with the launching of the National Biotechnology Strategy (NBS). This early strategy focused on "R&D and human resources development" rather than regulation (Government of Canada, Industry Canada, 1998:4). In 1988, the Canadian Environmental Protection Act was passed, and questions were raised at the time whether special provisions for the regulation of biotechnology should be made (Bjorkquist, 1999:19). Instead, the decision was made to amend a number of federal statutes to accommodate the new technology, in a manner similar to the American approach. Leiss and Tyshenko (2001) argue that a "business as usual" attitude to biotechnology regulation (i.e., incorporating it under existing legislation) was made possible by conceptualizing biotechnology with a "very broad definition" (324–325).

Consequently, in 1993 the Regulatory Framework for Biotechnology was announced. The regulatory framework had six key principles, which were to be "the basis of a federal regulatory framework for biotechnology" (Government of Canada, Canadian Food Inspection Agency [CFIA], "Federal Government"):

- maintains Canada's high standards for the protection of the health of workers, the general public and the environment;

- uses existing legislation and regulatory institutions to clarify responsibilities and avoid duplication;
- continues to develop clear guidelines for evaluating products of biotechnology which are in harmony with national priorities and international standards;
- provides for a sound scientific database on which to assess risk and evaluate products;
- ensures both the development and enforcement of Canadian biotechnology regulations are open and include consultation; and
- contributes to the prosperity and well-being of Canadians by fostering a favourable climate for investment, development and adoption of sustainable Canadian biotechnology products and processes. (ibid.)

Thus, a notable similarity to the U.S. framework is the Canadian move to regulate under existing statutes. Unlike the U.S. framework, however, which contains almost 50 pages detailing definitions and designating regulatory authority under various acts, the Canadian framework exists in little more than its news release and a question-and-answer document highlighting the six key principles. It is seemingly a shadow of its American counterpart. In 1997 a renewal process for the National Biotechnology Strategy was launched, the end result of which was the 1998 Canadian Biotechnology Strategy. Under this strategy, the government of Canada's commitment to the 1993 framework was reaffirmed, and it remains the basis of biotechnology regulation in Canada today.

Canadian biotechnology regulatory oversight is currently carried out by three main agencies: Health Canada, the Canadian Food Inspection Agency [CFIA], and Environment Canada. The CFIA is the lead agency for biotechnology regulation, and it is responsible for fertilizers, feeds, and seeds. Environment Canada serves an umbrella function, assuring that environmental assessments of health and environmental impacts are made. Health Canada is responsible for food safety. As in the United States, there is some regulatory overlap between agencies. For example, Environment Canada is responsible for environmental risk assessments of new substances with respect to toxicity unless another agency performs this assessment. For GM products, such assessments are performed by the CFIA. Health Canada and CFIA assessments are conducted according to the dictates of each agency, and authorizations are granted independently. The Canadian system is again a composite of "product-based" as opposed to "process-based" assessments.

Genetically modified plants are regulated by the Plant Biosafety Office of the CFIA under their designation as plants with novel traits [PNTs]. Such plants "may be produced by conventional breeding, mutagenesis, or more commonly, by recombinant DNA techniques" (ibid.). The CFIA is responsible for the importation, environmental release, and registration of PNTs, and so serves a function similar to that of APHIS in the United States. The CFIA monitors field trials, which are subject to "reproductive isolation." This denotes "conditions that mitigate the transfer of pollen," not necessarily contained settings (such as a greenhouse or laboratory) (Canadian Biotechnology Advisory Committee [CBAC], 2002: n.p.). The CFIA reportedly monitors the trial site over the years while the technology developer collects agronomic and environmental impact data (ibid.). If the developer believes the product has commercial potential, it provides a scientific information package in its application for unconfined release into the environment and marketplace (ibid.). The CFIA makes its assessment based on this data, and, if no risks are apparent, will authorize release.

In 1998, the Canadian government created the Canadian Biotechnology Advisory Committee [CBAC] in order to provide it with independent advice on biotechnology policy issues. While the CBAC claims that Canada is the "only country where regulatory oversight is triggered by 'novelty'" (ibid.), evidence of the same U.S. concept of substantial equivalence is readily apparent throughout the Canadian regulatory framework. For example, with respect to Health Canada:

> The basis of Health Canada's safety assessment process is the principle that novel foods can be compared with traditional foods that have an established history of safe use, and that this comparison can be based on an examination of the same risk factors that have been established for the counterpart food. (ibid.)

The concept is also explicit in the CFIA mandate, where it soon becomes clear that, practically speaking, "novelty" and "substantial equivalence" operate in a strained conceptual association. While PNTs are "not considered substantially equivalent" to other plants of the same species, the inclusion of non-GM forms of plant breeding under PNTs necessarily associates the regulation of GM plants with those of conventional breeding. Lastly, as in the United States, the concept of substantial equivalence extends to the marketing of the product. Once commercialized, the labeling of GM products in Canada is permissible, but voluntary (see Government of Canada, Canadian General Standards Board, 2004).

Pro-Development in the North American Bloc

In short, while there are differences between the Canadian and U.S. regulatory systems for agricultural biotechnologies, there are deep similarities. Both nations' federal regulatory frameworks use existing legislation, subjecting GM products to a patchwork of regulatory agencies and statutes, some overlapping. Both are "product" rather than "process" oriented, allowing no special provisions for the method by which GM products are produced. Both are based on the concept of substantial equivalence, whereby GM products are not seen to differ from those of conventional breeding. Both rely on data submitted by the product's own manufacturers for their safety assessments. Both conduct their assessments on a case-by-case basis and appear to address the potential uncertainty of this approach with a fair amount of agency-industry consultation. Both subscribe only to voluntary GM labeling.

Informal consultation and harmonization has obviously influenced Canada-U.S. regulatory convergence. There has also been at least one formal attempt at regulatory harmonization. In 1998, the CFIA, Health Canada, and the USDA participated in bilateral discussions "to compare and harmonize where possible, the molecular genetic characterization components of the regulatory review process for transgenic plants" and to "prioritize future areas of cooperation and information exchange" (Government of Canada, CFIA, Health Canada, and USDA, 1998). Whether formally or informally converged, these regulatory frameworks have also been similarly subjected to significant criticisms. Critics argue that incorporating agricultural biotechnologies under existing legislation does not allow for their unique aspects to be regulated (which fall into the gaps between regulatory authorities designed for conventional crops); that regulation based on the concept of substantial equivalence is inappropriate; and that both governments are so pro-development that they favor weak regulation. Numerous "incidents" have also provided physical backing to concerns over regulatory laxity.

In Canada, prolonged public pressure prompted the three key regulatory agencies for biotechnology to commission an independent review of their regulatory system by the Royal Society of Canada. In its 2001 report, the Royal Society advocated a more precautionary regulatory approach and made 33 recommendations toward strengthening Canada's regulatory standards. Notably, it suggested the need for independent auditing of the scientific and ethical aspects of biotechnology regulation and raised serious concerns about science-based risk assessment due to three factors: (1) the "conflict of interest" created by regulatory agencies' double mandate of promoting and regulating the industry; (2) the "barriers of confidentiality that compromise the transparency and openness to scientific peer review of the science upon which regulatory decisions are based"; and (3) the "extensive and growing conflicts of interest within the scientific community due to entrepreneurial interests" and the increasing corporate "domination of the research agenda" (Royal Society of Canada, 2001:ix).

The Royal Society further concluded that the CFIA's framing of substantial equivalence "links it intimately with the definition of 'novel trait' in a way that leads to a logical impasse" (ibid.:181). More significantly, with respect to a regulatory approach based on this concept, the Royal Society found that "the use of 'substantial equivalence' as a decision threshold tool to exempt GM agricultural products from rigorous scientific assessments to be scientifically unjustifiable and inconsistent with precautionary regulation of the technology" (ibid.:ix).

Faced with this report, the Government of Canada proposed an action plan that included a commitment to ongoing assessment of progress according to the Royal Society recommendations and in conjunction with advice from the CBAC's upcoming 2002 report (CBAC, 2002). In 2004, however, the CBAC issued an advisory memorandum warning that the "lack of a comprehensive regulatory system for products of biotechnology is impeding the development of niche industries in Canada and consequently the potential for consumer and economic benefits" (CBAC, 2004: n.p.). The memorandum stated that despite the CBAC and Royal Society reviews, "there is little evidence of government action to implement recommended improvements." Rather, despite five separate regulatory review and development processes, "not one of these efforts has delivered even draft regulations" according to original timelines: "In fact, there seems to be a practice of simply extending the target dates to some never quite attainable date in the future" (ibid.).

Consequently, despite a wealth of reports and communication, regulatory reform in Canada appears to be a low priority. Similarly, relatively little U.S. federal legislative activity on biotechnology has occurred. Overall, both governments indicate a strong concern with facilitating biotechnology development and with preventing regulatory hurdles from impeding that development. For example, the lack of independent data collection for crop assessments—which relies on manufacturers to be forthright about safety limitations of their GM products—could lead to insufficient data and testing, if nothing worse. At the very least, it suggests a significant level of cooperation between industry and regulators in facilitating industry development. In fact, both the Canadian and U.S. regulatory systems demonstrate a lack of separation between government as promoter and as regulator, and evidence of the keen interest in promotion can be found on both sides of the border. In Canada, industry "success" was early on associated with its development:

> The dramatic growth in biotechnology activity in Canada—from a small core of health care companies, to a community of more than 500 firms employing more than 25,000 Canadians, underscores the success of the National Biotechnology Strategy over the past 15 years. (Government of Canada, Industry Canada, 1998:4–5)

The 1998 renewed Canadian Biotechnology Strategy asserts the government of Canada's consistent "support for biotechnology as a priority" (ibid.:4). Indeed, its ten key themes emphasize measures to enhance Canadian competitiveness, such as through developing sectoral support, expanding research and development, and accelerating the application and commercialization of new technologies. Efforts have been made to dispel the impression of a conflict of interest. For example, Agriculture Canada was initially the "lead developer, promoter and regulator" of agricultural biotechnology until its regulatory function was transferred to the CFIA (Bjorkquist, 1999:25), but criticisms of the Canadian government's contradictory role as industry facilitator and regulator have continued.

This perceived conflict of interest came to a head over the issue of Monsanto's application to introduce Roundup Ready wheat in Canada in 2002, discussed in greater depth in Chapter 4. The salient point here is that the government had to assess Monsanto's application against a backdrop of intense and widespread opposition. It was ultimately revealed, however, that the government had a role in developing the technology (providing funds, experimental fields, and government scientists), and even stood to profit from a royalty of up to 5% of sales if the wheat was approved (CBC Television, 2003).

The promotional stance taken by the U.S. government is even more explicit, and given the country's international influence it also has greater global significance. For example, the 1986 "Coordinated Framework" subscribed to the goals of the Organization for Economic Co-Operation and Development report on recombinant DNA applications; indeed, the United States was a major contributing member to the report. Its seven general recommendations emphasize pro-development goals such as information sharing, harmonization, protecting intellectual property, and balancing "adequate review and control" with "avoiding undue burdens" (US, FR, OSTP, 1986:23308). That is, "any approach to implementing guidelines should not impede future developments in rDNA technology." The report further posits that "international harmonization should recognize this need" (ibid.). The United States is similarly active from a pro-biotechnology development stance on other international agreements. For example, despite polls indicating the vast majority of Americans are in favor of GM food labeling, the United States is actively campaigning at the UN Codex Alimentarius commission in opposition to mandatory labeling (Environmental News Service [ENS], 2006).

There has also been significant concrete evidence of regulatory failure in the two countries. In Canada, regulatory weakness is highlighted with respect to the technologies' proprietary aspects, as well as the legal altercations that have occurred in the absence of legislation governing liability for the involuntary presence of patented GM material. This came to a dramatic head in the Schmeiser case and continued in the Saskatchewan Organic Directorate–supported Hoffman case, to be discussed presently. In the United States, lawsuits over patent infringement through seed saving are mushrooming. Thus the number of unresolved proprietary issues and the growing number of lawsuits over GM crops indicate the need for legislation; however, regulators on both sides of the border appear reluctant to create it. While the courts will decide any given case, it will be based on the specifics of the claim in question, not on the broader social issues which legislation could address. Thus the reluctance to regulate is a reluctance to proactively assign the new distribution of power between farmers and biotechnology developers, among other issues. Unfortunately, as Galanter (1994 [1974]) suggests, the financial and power differential between farmers and biotechnology developers will be reflected in a similar imbalance in influence on any rule making that occurs as an outcome of litigation.

While Canada highlights regulatory gaps with respect to agricultural biotechnologies' proprietary aspects, numerous contamination "incidents" highlight serious flaws in U.S. biotechnology regulation and monitoring. As the United States has eight times the GM hectares as Canada, it is not surprising that it has more physical evidence of regulatory failure. Given that the two countries' regulatory frameworks are themselves substantially equivalent, however, there is no reason to believe the failures are not equivalent as well. A case in point is the 2009 GM contamination of Canadian flax: the GM flax was developed in 1990, but never commercialized, and all seeds were to be destroyed in 2001. But in 2009, it was found in Canadian fields, causing flax prices to plunge from $11 to under $3 a bushel (Mittelstaedt, 2009).

The most spectacular illustration of U.S. regulatory weakness was the Starlink Corn debacle. Production of Starlink Corn was approved for animal feed and industrial uses in 1998, but not for human consumption due to lingering questions about its allergenicity. Regulatory approval was conditional on a number of special procedures designed to keep it out of the human food chain—for example, mandatory segregation, buffer zones, and grower education; however, there was an apparent widespread failure to follow these procedures (Mandel, 2004). In 2000, traces of the unapproved corn were found in tacos and other corn products—significantly, not by regulators but by an environmental coalition. Starlink was regulatory failure writ large: "one company, with one GM crop, managed to contaminate food for millions of households and brought an international commodities market to a standstill" (Bratspies, 2003:593; see also Bratspies, 2002).

Starlink vastly increased public pressure on U.S. regulatory bodies to ensure the safety of biotechnology regulation. Subsequently, the EPA committed to not issuing any further split approvals. While there seems to be no written policy to this effect, Canadian officials likewise claim not to issue split approvals (CFIA, personal communication, 2006). Starlink showed the dangerous regulatory lapses that can result from the piecemeal approach to regulation common to both these countries. Moreover, it highlighted further gaps regarding liability for agricultural biotechnologies. According to Hamilton (2005), seed companies initially attempted to "allocate costs and liability to the 'offending' farmers, many of whom had never seen the restrictive terms of the product approval" (48). He contends that if an Iowa attorney general's office hadn't stepped in, "the whole episode may have evolved quite differently" (ibid.). Such attempts to escape liability are likely to be even greater in more ambiguous cases, as biotechnology companies attempt to shift responsibility onto producers through the technology transfer agreement "by placing language in [them] to make producers responsible for post-harvest 'channelling'" (48–49).

Despite a wealth of safety assurances since Starlink, regulatory failure persists. In 2002, two incidents of improperly contained pharmaceuticals led to the destruction of significant amounts of potentially contaminated crop (Mandel, 2004). In 2005, the Swiss company Syngenta inadvertently mixed its approved Bt11 corn with the similar but unapproved Bt10 corn, resulting in 37,000 acres of the unapproved corn planted in the United States, a portion of which was exported to the European Union (Wright, 2005). In 2006, GM LL601 rice, unapproved for human consumption in any country, was discovered in the United States, disrupting export markets and causing millions of dollars in damage to the U.S. rice industry (Union of Concerned Scientists, n.d., "Rice 'Mystery'"). In 2011, GM bentgrass that escaped from an Idaho trial site decommissioned in 2006 continued to spread through Oregon (Lies, 2011). Chemicals suitable for containing the glyphosate-resistant grass are unsuitable for use in waterways, where the grass grows best (Meyer, 2011).

In 2002, APHIS created the Biotechnology Regulatory Service [BRS]. According to the BRS website, its purpose is to place "increased emphasis on our regulatory responsibilities," although subsequent web pages report the familiar mixed purpose of regulation and development: "to focus on the USDA's key role in regulating and facilitating biotechnology" (APHIS website). This is likely why the BRS puts a strong emphasis on public communication about regulatory strength, while still presiding over regulatory failure. For example, according to the BRS, there are "serious penalties" for failure to adhere to BRS regulations, permit conditions, and requirements; depending on the crop involved, "a site may be inspected at least 5 times" to ensure compliance; and APHIS field tests find 98% regulatory compliance (APHIS website). To the contrary, two audits a decade apart by the USDA Office of the Inspector General [OIG] find that "the Department's efforts to regulate those crops have not kept pace" (USDA, OIG, 2005:i). According to the OIG: "Although APHIS agreed to improve its tracking of inspection reports following an Office of Inspector General (OIG) audit more than 10 years ago, the agency continued to lack an effective, comprehensive management information system to account for all inspections and their outcomes" (iii).

The 2005 OIG report characterizes a lax, uncoordinated and downright ineffective system. In some instances, a lack of necessary regulation and guidance was at issue: notification protocols were not reviewed and were often only provided verbally; reporting of final dispensation of pharmaceutical and industrial harvests was not required (consequently, "two large harvests of GM pharmaceutical crops remained in storage at the field test sites for over a year without APHIS knowledge" [ii]); and an applicant planted "regulated edible GM crops in an open field, where they were accessible to the public" (7) and could be eaten by passers-by. Most significantly, APHIS regularly lacked basic information about the location of field tests, as it didn't "consistently collect precise location information" (14). Even when APHIS knew test site locations, there was a lack of coordination between management and inspection units, such that "BRS has no assurance that the highest risk field sites are inspected" (iii).

Regulatory incompetence and laxity were not the only issue. Notably, permit sites were not inspected with the frequency APHIS reported, and the agency "understated to the public the percentage of inspected sites with compliance infractions" (7). Further, the OIG "found 11 violations that were not recorded in BRS' compliance infractions database," despite having been reported or being identifiable from information BRS already had, with administrative action taken on "only 1 of those 11 violations" (iii). Thus it would appear that in the dual pressures to regulate and facilitate the development of the biotechnology industry, development has far more weight in both Canada and the United States.

The European Union: A Precautionary Approach

Biotechnology regulation in the European Union stands in marked contrast to the pro-development bloc evolving in North America — a contrast that may have important consequences for the global dissemination of biotechnologies. The EU only took its current shape as a single-market economic union in 1992. Biotechnology regulation in the EU is doubly challenging because regulating this controversial new technology requires consensus from its member states. The challenge is ongoing and is exacerbated by the EU's expanding membership.

In 1990, the EU adopted Directive 90/220/EC for the assessment and approval of GM organisms, and subsequently it approved a number of GM products for commercial use, including crops. In 1997, Regulation 258/97/EC was passed, requiring the labeling of foods containing GMOs, excepting foods that were derived from GMOs but did not contain them (Pew, 2005:9). Biotechnology regulation in the EU is affected by a greater cultural sensitivity toward food, and concerns have been exacerbated by food safety incidents, such as the bovine spongiform encephalopathy (most commonly known as BSE) outbreak in the mid-1990s. Perhaps due to this greater sensitivity, it has proved difficult to obtain member state consensus about biotechnologies. Starting in 1997, a number of member states invoked the "safeguard clause"[1] to ban crops from their countries that had already gained EU approval. Many of these later vowed to "block approval of GM crops unless existing labelling and safety regulations were further tightened" (ibid.:10). This resistance resulted in a de facto moratorium, where no new approvals were granted while the EU worked to develop legislation that would satisfy its diverse member states.

In 2003, while these regulatory amendments were ongoing, the United States, Canada, and Argentina—top biotechnology crop producers—launched a WTO challenge over the EU's de facto moratorium. They argued that it constituted an unfair trade barrier and thus violated free-trade agreements. Further, they believed a challenge was "necessary to discourage other countries, especially those in the developing world, from using the EU regulatory approach as the basis for their own regulations" (ibid.:12). By 2004, when the de facto moratorium was lifted, the EU had, in fact, produced a number of relevant legislative amendments and new initiatives. The general theme of this new GMO legislation, in contrast to the North American model, was a pronounced emphasis on detailed information gathering, risk assessment, and tracking and monitoring procedures. Also, in contrast to the North American model, GMOs are not regulated under a compilation of existing statutes, but according to legislation specifically crafted for the purpose.

In the EU, authorization for experimental release of GMOs is a national decision, based on EU risk assessment procedures. An EU member state may grant authorization based on an evaluation of environmental and health risks outlined in the application and on conditions laid down in Directive 2001/18/EC. Unlike APHIS' notification system, which lacks even basic siting information on such high-risk crops as pharmaceuticals, the EU application package requests detailed information not only on siting but also on data relevant to assessing potential impact on the "receiving environment," such as proximity to humans; proximity to drinking water supplies; nearby flora, fauna, livestock, and migratory species; and potential future land developments, among others. The application must also include detailed information on monitoring plans, procedures for controlling spread, waste treatment, and emergency response in case of unexpected spread (Official Journal of the European Union [OJEU], 2001: Annex III A:III).

While field trials are a national decision, authorization to market GMOs includes all nations: under the EU common market, commercialization of a GMO in one nation implies authorization in all. Application for authorization is presented to one member state, which produces an assessment report based on the risk assessments of the field trials and on further assessments of information specific to marketing the product, such as intended use, storage, and handling, and identifiers for post-marketing control and inspection (ibid.: Annex IV). If the member state's assessment is favorable, the application is referred to the European Commission. If there are no objections from other member states or the European Commission, the GMO can be marketed (Europa, 2005b:7).

New regulations put in place in 2003 (Regulation 1829/2003) allow for even greater centralization of authority. Manufacturers can apply for authorization under a two-part assessment (including Directive 2001/18/EC and Regulation 1829/2003, specifying new criteria for GMO food and feed) or they can file a single application under 1829/2003 for authorization for all uses of a GMO (both environmental release and use in food and feed). The latter option streamlines the process according to a "one door, one key" principle (Europa, 2005b:10). Under 1829/2003 an application is assessed by a single risk assessment conducted by the European Food Safety Authority and a single risk management process involving commission and member states in a regulatory committee procedure (ibid.). The Food Safety Authority is an independent source for scientific advice and technical support, endowed with legal personality (Europa, 2005c). Unlike the Canadian Biotechnology Advisory Commission, the European authority is responsible for disseminating science knowledge, without broader policy concerns. Its mandate includes providing scientific opinions, developing uniform risk assessment methods, commissioning scientific studies, identifying and characterizing emerging risks, collecting and analyzing food safety data, and compiling an inventory of European-level data collection systems (ibid.).

The European model thus has a much greater potential for science-based regulations than the "don't look, don't tell" American approach. Overall, EU agricultural biotechnology regulation provides a number of significant contrasts to Canada and the United States. The EU itself notes that the "one door, one key" approach ensures that "experiences such as with Starlink maize in the US . . . are avoided because GMOs likely to be used as food and feed can only be authorized for both uses" (Europa, 2005b:10). Further differences with respect to regulations on traceability and labeling, and with respect to co-existence and liability issues, mark an even greater policy divergence.

In 2003, the EU passed Directive 1830/2003, which put traceability and labeling requirements into place for all member states, effective in 2004. The labeling requirements specify that prepackaged products must contain labels stating if the product contains GM organisms and that products that are not prepackaged be displayed in association with such labeling. The traceability legislation requires that from the first stage of placement on the market (including in bulk), and all subsequent stages, written records must be kept (and maintained for five years) indicating that the product contains GMOs and providing the unique identifiers of those GMOs (Official Journal of the European Union [OJEU], 2003: Art. 4:4). Exemption thresholds for these requirements were put in place to allow for trace amounts of "adventitious or technically unavoidable" GMOs (ibid., Art. 4:7). The new regulations reduced this threshold to 0.9 percent (from 1.0 percent) and extended its application from only those products that tested above 0.9 percent to all products derived from ingredients that were above 0.9 percent GM (Gene Watch UK, n.d.). This allows for products such as vegetable oil, which are so refined that GM material is no longer detectable, to be included in the labeling requirements if they were produced from GM ingredients.

While labeling is about consumer choice, traceability requirements provide the ability for future monitoring as well as a contingency plan in the event that a GM product turns out not to be as safe as its conventional counterpart. Directive 1830/2003 states:

> Traceability requirements for GMOs should facilitate both the withdrawal of products where unforeseen adverse effects on human health, animal health or the environment, including ecosystems, are established, and the targeting of monitoring to examine potential effects on, in particular, the environment. (ibid.:2)

These requirements thus offer a significantly more stringent regulatory environment than in North America.

Again in contrast to Canada and the United States, the EU has further attempted to directly address the issues of co-existence and liability. In 2003, the commission decided that member states could develop their own measures regarding co-existence. It produced guidelines for the development of such measures based on the principle that "farmers should be able to choose the production type they prefer, without forcing them to change patterns already established in the area," and that, in general, "farmers who introduce the new production type should bear the responsibility of implementing the actions necessary to limit admixture" (Europa, 2005b: 17).

Consequently, a number of EU states worked on co-existence legislation, which can include such elements as separation distances, buffer strips, monitoring, and requirements to inform neighbors of GM crops. Member states can also dictate who bears the financial burden in case such procedures are insufficient to contain the spread of GM material.

German co-existence legislation, for example, puts in place a "polluter pays" principle, such that GM farmers would be liable for contamination to non-GM farmers' fields. This liability could be joint where individual liability is not possible to determine (Reuters, 2005). Denmark developed a scheme whereby compensation (funded by a tax on GM producers) will be paid to conventional or organic farmers who suffer economic loss due to GM contamination in their crops (Europa, 2005a). In the Netherlands, the Agriculture Ministry urged agricultural stakeholders to construct a voluntary agreement on cohabitation. Pro-biotechnology groups were motivated to do so given that GM crops had not yet been commercialized there. An agreement was ultimately crafted which again included a special compensation fund in cases of contamination (Mudeva, 2004). The generation of such liability legislation avoids the huge regulatory gray area evident in Canada and the United States.

Subsequent to the new labeling and traceability requirements, the EU ended its de facto moratorium with the approval of Syngenta's GM sweet corn. Since 2004, a number of GM products have been approved. A homogenous approach to biotechnology regulation in the EU was still not assured, however. Despite the new legislation, the unpopularity of GMOs with the European public prompted some countries to maintain their bans on approved GM crops using the "safeguard clause." The commission called on dissenting nations to lift their bans, but its recommendation "to force the lifting of the national bans was rejected by a qualified majority of the Council" (Pew, 2005:16). Ultimately, ongoing member state–EU tussles over GMO approvals forced the EU to propose an amendment to Directive 2001/18/EC which would give member states full responsibility for GM cultivation in their territories (Europa, 2010). Thus it would allow them to account for "local, regional and national conditions" and to "restrict or prohibit the cultivation of GMOs in their territory" without recourse to the safeguard clause (ibid.).

In short, EU biotechnology regulation markedly differs from the North American model. It emphasizes data gathering, tracking, and long-term monitoring. Almost antithetically from North America, the emphasis is on visibility in decision making, long-term assessment, and public consultation. While GM labeling of consumer products is currently limiting market opportunities for them, the clarity and single-desk authorization process may provide some corporate benefit. The contrast with North America is not coincidental. A 2002 European Commission strategy paper for biotechnology stated that the technology was developing so rapidly that Europe's policy choice is "not whether, but how to deal with the challenges posed" (Commission of the European Communities,

2002:4). Consequently, it emphasized the need to make these choices before they were imposed: "Europe is currently at a crossroads: we need to actively develop responsible policies in a forward looking and global perspective, or we will be confronted by policies shaped by others, in Europe and globally" (ibid.).

While the United States' drive to promote biotechnology, in conjunction with international laws, is applying some pressure on the EU, the EU is nonetheless developing its own terms for biotechnology adoption and dissemination.

Resistance and Inevitability in the Contours of the Third Food Regime: Global Politics, Local Resistance

The prevailing image of biotechnology development in Canada and the United States is one of America as biotech superpower, with Canada running tag-along development. Together, these two nations form an industrialized bloc promoting corporate biotechnology. The EU demonstrates a countervailing tendency. While it also supports its biotechnology industry, it emphasizes a much more precautionary form to development. Developing countries form a third bloc (albeit with differentiation), both with respect to their position as a market for the technologies and as a market for the surplus produced with these technologies by developed countries. Thus they are an important component of the global food regime and an increasing factor in biotechnology power relations between the United States and the EU:

> A main objective must be to ensure that the EU maintains competitiveness vis-à-vis major industrialised countries such as the US and Japan. Moreover, whatever policies Europe will decide regarding life sciences and biotechnology, they will have important international impacts, in particular for developing countries. . . . [W]e need to develop an international agenda, based on our fundamental values and long-term objectives, to actively promote balanced and responsible policies globally, in particular towards the developing world. (Commission of the European Communities, 2002:25)

Not to overstate the importance of the EU's biotechnology industry: the U.S. biotechnology industry "started earlier, produces more than three times the revenues of the European industry, employs many more people (162,000 against 61,000), is much more strongly capitalized and . . . has many more products in the pipeline" (ibid.: 9). Nonetheless, the EU's antithetical stance to American-style biotechnology has an influence on the global stage. Some, like Bernauer (2005), argue that this influence is an important factor in the United States' WTO challenge, as the EU could otherwise influence broader geopolitical relations around biotechnology:

> Both the US and the EU have, in recent years, been negotiating free-trade agreements with developing countries. Those agreements, some of which also cover non-tariff barriers to trade, such as environmental and consumer risk regulation, may have the effect of locking in either the US or the EU model of agri-biotech regulation in large parts of the developing world. (Bernauer, 2005:24)

Consequently, the United States has some motivation to challenge the EU and deter developing countries from "emulating the EU's agri-biotech regulation," potentially compromising U.S. market access for its GM products (ibid.).

While the broad contours of an American-dominated neoliberal food regime are already apparent, there are still significant factors that could upset it. Three appear likely to have the greatest influence: (1) the development of international agreements and (equally significantly) the power that nations attribute to them; (2) the impact of shifting national priorities resulting from domestic pressures, such as legal changes or significant subnational (or sub-union, in the case of the EU) resistance; and (3) the costs and benefits of the technology itself, as the technology is still sufficiently new that dramatic developments could shift global opinion—in either direction. Notwithstanding the latter's importance, the discussion here will focus on the first two.

Food regime scholars such as McMichael argue that under international trade agreements such as GATT, "national sovereignty would be subordinated to an abstract principle of membership in the state system that sanctions corporate rights of free trade and investment access" (1992:354). Consequent constraints on national autonomy result from both international agreements and the internationalization of capital itself. State subordination under the power of internationalized capital does not yet seem to have been a factor in biotechnology regulation. As biotechnology is increasingly integrated into the global food regime, however, nation-states will likely become more vulnerable to its dictates.

There are already indications that international agreements such as those which recognize intellectual property rights could be an important constraint. Further, as the United States is frequently a central force in the development of these agreements, they are likely to facilitate consolidating American power.

Vandana Shiva (2001) argues that the United States aimed to make intellectual property "its primary asset for economic growth" (19) and that a global patent system was central to this aim. The adoption of the TRIPS Agreement imposed just such a system and secured U.S. advantage over developing countries with respect to patents on life. As noted, TRIPS requires countries to either adopt a patent or sui generis system of plant variety protection, such as UPOV. However, UPOV is heavily weighted in favor of protecting corporate rights over national social policy goals. The homogenizing effect of TRIPS is not yet certain, as some countries are resisting the U.S. version of plant variety protection. Further, the ambiguity with respect to how TRIPS reconciles with the Convention on Biological Diversity has led to some heated debate with respect to which has precedence (ibid.:103). Some even argue that the CBD and the Cartagena Protocol "have become staging grounds for resistance to WTO rules and to the market-based management of genetic resources that the WTO supports" (McAfee, 2003:175–176). Battles are also waged in the WTO and associated agreements. Two examples are U.S. efforts to impose voluntary (rather than mandatory) labeling at the Codex Alimentarius and its WTO challenge of the lawfulness of EU labeling requirements (Davidson College, 2004).

In sum, the most significant factor shaping the food regime will likely be the nature and force of international agreements, *where nations are sufficiently motivated to abide by them*. For example, the punitive countermeasures predicted to come out of an EU loss at the WTO (Bernauer, 2005) would only affect the EU if the EU found it politically expedient to comply. In 2006, the WTO indeed ruled that the EU's de facto moratorium was illegal, although the EU already had its new agricultural biotechnology regulations in place and had already lifted its moratorium. Thus the ruling had more political than practical impact. Tellingly, even prior to the WTO ruling, the European Commission stated that the WTO report would not force changes to EU regulations: rather, it emphasized, "[o]nly products recognized as safe will be allowed and the WTO report will not influence the decision-making process in the EU" (cited in Minder, 2006).

Notwithstanding the potential constraints of international regulations (where there is sufficient motivation to abide by them), the three regions here have demonstrated a great deal of national regulatory autonomy. For the time being, agricultural biotechnology *is* a national or EU-driven enterprise, and whatever happens within these political entities to affect their priorities can affect the shape of the evolving regime. These priorities can change due to political pressure from below. This is most evident in the EU, with its historical trouble getting member state compliance over GMOs. As concluded by Kurzer and Cooper (2007), "[EU] countries with hostile publics tend to turn down requests for permission to sell transgenic seeds, crops, or foods, whereas countries with tolerant publics tend to approve requests" (1052). Such bottom-up pressure is not unique to the EU, however, but is evident to a lesser extent in both the United States and Canada.

One study of U.S. legislative activity on agricultural biotechnology, for example, found that while its regulation "was not a top priority for Congressional legislators," states "have emerged as the key battlegrounds of issues raised by agricultural biotechnologies" (Pew, 2003: n.p.). Between 2000 and 2001, 158 pieces of legislation were introduced at the state level. While many of the bills and resolutions dealt with preventing crop destruction (28%), a number dealt with issues such as regulating GM crops (12%), liability and agricultural contracts (18%), and labeling GM foods (16%). Of course, the majority of initiatives did not pass, and out of those that did, a full 67% dealt with GM crop destruction. This activity is nonetheless highly significant and is broad based (initiatives were from 39 states), indicating bottom-up pressures on the U.S. government over GMOs.

Local-level initiatives and litigation abound. An initiative that struck a similar chord to the policy perspective of the European Union, for example, was the failed 2002 initiative in Oregon that proposed labeling of GM food. Another tactic has been to sue the USDA over the failure to obtain Environmental Impact Statements for released GM crops. This tactic has forestalled the commercialization of GM alfalfa in both the United States and Canada (see Chapter 5) and is currently creating significant problems for GM beets in the United States (Voosen, 2010). Most recently, in March 2011, the Public Patent Foundation (PUBPAT) launched a suit in the New York district court challenging Monsanto's patents (*Organic Seed Growers and Trade Association et al. v. Monsanto Co. et al*). Their amended complaint had 83 plaintiffs, including farmers and seed businesses (mainly organic), concerned that they would be involuntarily contaminated with GMOs and then accused of patent infringement. Their claim sought preemptive assurance that this would not happen.

In addition, numerous regions of different sizes in the United States and Canada have hosted petitions for GMO-free zones. In Canada, Powell River, BC, became the first GM-free crop zone in 2004, and an initiative was put forward to make Prince Edward Island GMO free, but did not pass. Also in 2004, Mendocino County, California, became the first U.S. county to ban the propagation of GM crops and animals. A spate of similar initiatives followed suit in California following its success, and a limited number—such as Trinity, Marin, and Santa Cruz counties—successfully passed. The forces of opposition are also not to be underestimated, however, and the legacy of anti-GM activities necessarily includes these countervailing forces (for more on the California initiatives, see Pechlaner, n.d.). One backlash to resistance in California, for example, has been a concentrated effort by industry and its proponents to pass state preemption legislation prohibiting local governments from taking such positions. A number of such laws have already been passed. In North Dakota, for example, Bill SB2277, passed on March 16, 2005, reads: "A political subdivision, including a home rule city or county, may not adopt or continue in effect any ordinance, resolution, initiative, or home rule charter regarding the registration, labeling, distribution, sale, handling, use, application, transportation, or disposal of seed."[2]

Such laws do not necessarily spell the end of political resistance, of course, but can temper its potential or change its course. While state preemption did not pass in California, for example, the resulting political turbulence led to California's Assembly Bill 541, passed in February 2007. Although the legislation did not contain all that the technologies' detractors hoped, it was the first legislation in North America to address liability issues around GM crops (Pechlaner, n.d.). Local activities consequently carry a potent political force if sufficiently cumulative. Thus while the contours of a U.S.-led neoliberal food regime appear set, the food regime, like globalization itself, is a contested phenomenon, and its trajectory "is conditioned by its resistances" (McMichael, 2004: n.p.). In Canada and the United States in particular, resistances are increasingly occurring in the legal forum.

Conclusion

The regulatory regime for agricultural biotechnology in the United States and Canada has been characterized by regulatory gaps, poor compliance, and other indications of weakness. Rebecca Bratspies argues the "central culprit" of this regulatory weakness in the United States "is a laissez-faire regulatory philosophy" (2003:631). It might be more accurately contended that the U.S. approach is less the result of passive oversight than it is an active program to facilitate the development of an American biotechnology empire. While agricultural production itself is no longer a driving force in industrialized economies, the input and processing sectors around agriculture provide vast opportunities for business to "farm the farmer." Agricultural biotechnologies play a central role in this business.

Following on the trends of the first and second food regimes, agricultural biotechnologies further the unfolding regime's integration tendencies. Based on global adoption rates, it is not too early to claim that they will likely be central to the next world food order. Corporations are central to biotechnologies, and the evidence to date suggests that the third regime will be an American-led neoliberal food regime, undergirded by corporate agricultural biotechnologies. Rather than this corporate rule presiding over powerless nation-states, however, both the United States and Canada provide strong evidence that regulatory weakness is more about state predisposition than lack of national autonomy. Thus the "corporate era" of agricultural biotechnologies and the neoliberal food regime is clearly a state-facilitated project: the key regulatory motivator appears to be ideology, not inability, and ideologies can be changed.

Such a finding needs to be qualified, given the significant power disadvantage of less well-developed countries. While GM crop adoption is increasing globally, it is increasing much faster in developing countries. A technology that can transform agriculture on this scale can provide massive economic benefits for its producers—most notably, the United States—much like the Green Revolution of the 1960s (Otero and Pechlaner, 2008). In this scenario, developing countries are most likely to be rendered political pawns, as already evidenced in United States and EU positioning over food aid shipments of GM grain to Africa. Yet, despite the United States' primacy in the industry, its continued pressure tactics, and WTO challenges, the EU, for one, has persisted in more precautionary biotechnology regulation, which itself is a challenge to the unfolding U.S.-style neoliberal food regime. Other players, such as Canada, could also forge a different path if they so chose, or if the domestic pressures were sufficient to motivate their doing so.

CHAPTER 3

Biotechnology on the Prairies:
The Rise of Canola . . .

We are in favor of the fact that they are bringing in new biochem products to market. . . . So if biotechnology can increase productivity, increase our net returns for farmers, we are in favor of it. We are in favor of anything that would increase our net profitability.
SK#21, SASKATCHEWAN CANOLA GROWERS ASSOCIATION

Introduction

Bordered by Manitoba on the east and Alberta on the west, Saskatchewan is at the heart of Canada's prairie region. The province occupies 161 million acres and has a population of almost 1 million (Government of Saskatchewan, "About"). The urban centers and the majority of Saskatchewan's population live in the prairie southern half of the province. To the north is the Canadian Shield, characterized by inaccessible rock and forested wilderness. In addition to its distinctive geography, climate and weather play a key role in Saskatchewan. Saskatchewan winters are long, cold, and dry, while summers are short, hot, and dry. Saskatchewan receives an extremely large amount of sunshine each year, gaining the designation of Canada's sunniest province, with some areas of the province even rivaling locales such as Rome in terms of hours of sunshine. Nonetheless, the short summers mean a relatively short growing season, with 60 to 100 annual frost-free days (International Society for Horticultural Science, "Canada").

Within the province itself, there are significant regional differences. The far north is severe and non-agricultural. The growing season increases as you go south, and the southern half of the province ranges from moderate to semi-arid conditions, such as in the American southwest. Even within a given region there is great seasonal variation: historical temperatures for Regina, in the southern third of the province, for example, have ranged from -50 to over 0 degrees Celsius in January, and from well below 0 to 35 degrees Celsius in July (International Society for Horticultural Science, "Canada"). Such variation makes weather a key factor in agricultural production, and makes drought and frost tolerance in crops traits of high value.

As of the 2006 Census of Agriculture, Saskatchewan has 44,329 farms, covering 64.3 million acres (Saskatchewan Agriculture and Food [SAF], 2007). Saskatchewan alone has 44% of Canada's cultivated farmland. The number of farms in the province is continuing its steady decline from 1936 (11.2% from 1996 to 2001; 12.4% from 2001 to 2006), which is consistent with a national trend of farm numbers decline since 1941 (ibid.). Farmland itself is only marginally declining, however, and still covers over 40% of Saskatchewan, with a vastly greater percentage in the more habitable southern half (Saskatchewan Agriculture, Food and Rural Revitalization [SAFRR], 2002a). In 2006, the average size of a Saskatchewan farm was 1,449 acres. This average incorporates a large number of hobby farms, however (SAF, 2007). In commercial agricultural production in Saskatchewan, 2,000 to 3,000 acres is reportedly more the norm.[1] Farms are also increasing in size: between 2001 and 2006, those from 2,880 to 3,510 acres increased 9%, while those 3,520 acres and over increased 27% (SAF, 2007). Despite agriculture's decline in significance (while the service sector increases), it nonetheless remains very important to the province's economy, generating about $2 billion annually (Statistics Canada, 2001) and employing 10% of its workforce (Government of Saskatchewan, Bureau of Statistics, 2004).

Biotechnology is another significant part of the province's economic plan. The provincial center of the bio-economy is Innovation Place — one of the province's two research parks — located at the University of Saskatchewan in Saskatoon. Innovation Place is responsible for 7,900 jobs; in 2004, it contributed $184 million to the economy of Saskatoon and $262 million to the economy of Saskatchewan (Government of Saskatchewan, 2005). The park focuses on agriculture, information technology, and the life sciences, with approximately 35 companies specifically engaged in agricultural biotechnology research and development. This accounts for about 30% of the nation's activity in this area, and according to the National Research Council of Canada [NRCC], it is "recognized as one of the largest clusters of its kind in the world" (NRCC, 2005: n.p.). It is here that the world's first GM commercial canola variety was developed (ibid.).

Both agriculture and agricultural biotechnology are therefore very significant to the provincial economy. At the same time, Saskatchewan has the largest acreage of organic production in the nation, making up 33% of all organic agriculture production. By 2003, there were 1,049 certified organic producers in Saskatchewan—over 2% of Saskatchewan's farmers—and this number is growing, with another 25 farms in transition to certified organic status (University of Saskatchewan, 2004). While relatively small in numbers, Saskatchewan's organic producers are nonetheless quite vocal on their issues. Biotechnology is one of these, not only because the philosophy of organic production is antithetical to the type of agricultural production that genetic engineering represents, but because GMOs are prohibited in organic production. Saskatchewan has also been the stage of two highly publicized lawsuits over genetic technologies that have formed the hub of some significant anti-biotech resistance. The impact of biotechnology on agriculture in the province is thus particularly interesting given this polarized context.

Saskatchewan farmers grow mainly grains, oilseeds, and specialty crops. The top five field crops in Saskatchewan are spring wheat, barley, alfalfa and alfalfa mixtures, canola, and Durum wheat. Thus wheat, of various types, is a highly important crop; its significance for agricultural biotechnology in Saskatchewan is illustrated in the following chapter. Producers also grow a variety of coarse grains, such as oats and rye; oilseeds such as flax; pulse crops, such as lentils and peas; and mustard and canary seed, among others. To date, transgenics have only been under significant commercial production in one crop in Saskatchewan—canola—although a very small area of transgenic corn is growing, and attempts have been made to introduce other transgenic crops in the past, such as flax and wheat.

Canola is an important crop to Saskatchewan. In 2001 it was produced on 33.4% of farms (Statistics Canada, 2004). Production patterns do differ by region, and many variations exist; however, a classical rotation pattern for a Saskatchewan farmer would be to grow an oilseed, then a cereal, a legume, and a different cereal or canola again, growing canola every third or fourth year. Because canola has been higher priced in the past, some growers try shorter rotations, some even growing canola every second year. In 2005, there were two kinds of GM canola on the market in Saskatchewan, Monsanto's Roundup Ready [RR] canola (resistant to its herbicide Roundup) and Bayer CropScience's Liberty Link [LL] canola (resistant to Liberty herbicide). Market share statistics are difficult to obtain, but knowledgeable sources estimate that 92% of the canola grown in Saskatchewan is herbicide tolerant of some sort. Of that, approximately 45% of canola is RR, 30% to 32% is LL, and 15% is Clearfield canola (SK#28, Saskatchewan Canola Development Commission). The Clearfield canola, by BASF, is an herbicide-tolerant canola that is produced through a process called mutagenesis, and is not genetically engineered.

Liberty Link canola, owned by Bayer CropScience, was the first GM herbicide-tolerant canola to be introduced. The first Liberty-resistant canola plants were open pollinated varieties, and were introduced in 1995. Monsanto first introduced RR canola, also an open pollinated variety, in 1996. Whether due to LL's poorer yields, as has been suggested, or due to producers' greater familiarity with Roundup, Monsanto's RR canola soon captured the bulk of the herbicide-tolerant canola market. In 1997, the Liberty-tolerant trait was introduced into a class of hybrid canolas called InVigor, which markedly improved yields. Currently, Liberty Link canola's market share is reportedly on the increase. At the same time, Monsanto is progressively shifting its Roundup-resistant trait to hybrid varieties.

Given my focus on the changes brought by the legal aspects of agricultural biotechnologies, interviews included litigants directly involved in the Schmeiser and Hoffman lawsuits, organic producers (given their attempt to certify as a class), conventional and GM producers, and stakeholder organizations in agricultural production. While producers who grew canola were of particular interest, wheat growers also had a vested interest in GMOs, given the attempted introduction of RR wheat in the early 2000s. This and the following chapter draw primarily on 34 of these interviews; 17 with producers (7 organic, 8 GM, and 2 conventional), 14 with representatives of agricultural organizations and other stakeholders (including government), many of whom were producers themselves, and 3 with knowledgeable informants. The remaining interviews in Saskatchewan relate more specifically to the litigation, and feature more prominently in Chapter 5.

Who Wants to Grow Genetically Modified Herbicide-Tolerant Canola?

With 92% of production, herbicide-tolerant canola is clearly the preferred approach for canola production in Saskatchewan, and within that, the GM varieties—with Roundup Ready and Liberty Link varieties together making up about 75% of production—are the most desirable. Much of the remaining 8% or so of canola that is not herbicide tolerant is selected for a specific purpose outside of what is offered in GM varieties. For example, there are two types of canola—Argentine canola (*Brassica napus*) and Polish canola (*Brassica rapa*)—but only *Brassica napus* is genetically modified. *Brassica rapa* is lower yielding, but requires a shorter time to maturity, and so it is favored in the north, where the growing season is shorter and the risk of frost before maturity is much greater. Only conventional breeding is done on *rapa*, and this explains a great deal of the conventional canola grown in Saskatchewan. Further, a number of specialty canolas—which are produced on contract for specific purposes, such as high-oleic-acid canola—are still predominantly conventional, although increasingly they are also being offered in herbicide-tolerant varieties.[2]

Opinions about GM canola in Saskatchewan vary; however, there is a general consensus about the technology among its users. For those who use GM herbicide-tolerant canola, the benefits have been firmly established in terms of improved weed management for a high-value crop. Canola is one of the more difficult crops to grow with respect to weed control, as it is more easily out-competed by them than other crops. There are also many weed species that are closely related to canola, so unlike some other crops, there was no good herbicide that could selectively treat weed problems in canola without damaging the crop itself. For this reason, prior to herbicide-tolerant varieties, canola could not be grown at all on land with significant weed pressure. The RR and LL systems allow for in-season weed control, and farmers can spray herbicide over the top of the already growing canola plants, drastically reducing weed pressure as an agronomic factor and allowing canola to be grown where it previously wasn't possible:

> One of the reasons I haven't grown canola over the years is weed control problems, and that's been solved by these new technologies. Well, with the chemicals we can use now we can get a lot better weed control in the crop. It's a lot simpler. (SK#14, GM producer)

We plant canola now where we would never think of planting it because of . . . it all comes down to weed management. There are farmers now who plant canola on just the roughest, dirtiest fields—dirty meaning lots of weeds—that would never have thought of planting it before. (K#1, GM producer)

Not only can GM canola now be grown where it previously couldn't, but the resulting canola is "cleaner," or freer of weeds, due to the post-emergence application of herbicide. This increases the farmer's yields (as fewer weeds mean less competition for water and nutrients) and increases the value of the crop (which now has less dockage—weed seeds, chaff, etc., mixed in it). High dockage can reduce the value of a crop significantly.

When we had the open pollinated varieties, we had problems with weeds that we couldn't kill out there, in the crop, they just grew uncontrolled out there and then you had more dockage. Whereas with these new varieties, these altered gene varieties, you can grow a cleaner crop. (SK#29, GM producer)

Herbicide-tolerant canola has also coincided with new perspectives on moisture retention. While leaving land to summer fallow had been the traditional philosophy of moisture preservation in Saskatchewan, the current agronomic philosophy is that this leaves soil vulnerable to erosion, and continuous cropping is advocated. Herbicide-tolerant canola leaves fields clean of weeds and facilitates direct seeding onto the previous year's crop stubble. The reduction in summer fallow provides farmers with an opportunity for increased income, as they can keep more land in production each year. Direct seeding also reduces the amount of cultivation, and thus the associated work time. This again provides more management flexibility, particularly for those who farm many acres or who have other constraints, such as working off farm. As one young farmer described it:

We were running out of time. My dad works out like me. We were both working. For us we just found we would have to be working the fields in the fall, preparing them for seeding, trying to get the weeds out—we just found we were limited with time, and the more land we took on the less time we had for prepping our fields, and in a way we just thought this would be a way in which it would make our lives a bit easier. (K#1, GM producer)

Given that GM canola allows producers to spray after the plant has already emerged, it also widens the chemical application time. As one crop development specialist characterized it:

> So somebody that is farming on the weekends, being able to spray Roundup or Liberty, and the crop is safe at any stage, that is pretty attractive. (SK#2, academic, crop development)

Therefore, in addition to weed control, GM technologies appear to provide some significant management benefits. It is not without management drawbacks as well, however, particularly with respect to RR technology.

One significant drawback results from the widespread use of Roundup as a general herbicide prior to the introduction of RR canola. Roundup had become the mainstay of farmers for spring burn-off and for other jobs related to weed management. Roundup Ready canola, therefore, added a management complication, as volunteer RR canola from a previous crop becomes a weed for any non-canola succeeding crop—one that cannot be managed by the usual treatment of Roundup or any glyphosate (the chemical in Roundup) herbicide. As one GM producer described it:

> Sometimes you don't just have volunteer canola the following year. . . . you might have canola show up in your field two years down the road, three years. A canola seed can stay in the ground for some time. So say three years down the road, you go out to spray your field in spring, burn it off with Roundup, and you see all the canola laying there. Right? And you want to sow flax on there. Well, you can't put 2-4-D in your tank to kill the canola because it's a residual, it'll kill your flax when it comes up. So that's a definite drawback, you really have to watch the following years. . . . You have to think about what you are doing more, you know. Plan your fields better, I guess. (SK#26, GM producer)

For most producers, this added management was well worth the benefits brought by the RR technology. For others, using LL canola had yield advantages as well as avoiding any of the management problems associated with the use of glyphosate-based herbicide tolerance.

Given the huge significance that health and environmental risk factors play in international debates about GM technologies, this might seem another area of importance to producers. Despite the fervor that swirls around these issues outside of the agricultural sector, however, I encountered few strong opinions on them at the farm level. As one producer explained it, while many producers may have some opinion on these debates, most just work out whether the technology will make things better for their farms, and base their usage decision on that. Even more pointedly, another producer characterized the necessary primacy of the bottom line:

> I think they [the health, safety, and environmental risks] are significant, but so far I don't believe that the concern in a mass way is preventing the primary producer from producing them. I think the pressures, the economic pressures, are so great on the farm that they're willing to accept those risks. (SK#11, GM producer)

Such expressions over broader environmental and health implications are rare outside of organic production, however. This is not surprising given the concerted effort of industry leaders to promote GM canola's safety—a message that, if compromised by producers, could hurt consumer demand. Chemical use, cross-pollinations, and weed resistance to chemicals (the "superweed" issue) are environmental issues that more directly overlap with production issues, although even these issues did not garner a great deal of concern from non-organic producers interviewed.

The perceived environmental benefits of using Roundup over other chemicals was occasionally raised:

> What I used to use as far as insecticides, fungicides, and herbicides . . . what I used to use and how lethal it was, compared to today, second to none. (SK#4, Canola Council of Canada)

This type of comment was most often heard from institutional organizations, but occasionally also by farmers. On the other hand, what farmers readily discussed, but industry and many organization leaders didn't, was the increased need to add chemicals to Roundup in order to manage RR canola volunteers. These chemical additions not only counter the proposed environmental benefits of using Roundup, but also add an additional cost to the overall chemical bill. The issue was particularly emphasized with respect to the attempted introduction of RR wheat, which would have resulted in two important crops requiring additional chemicals in order to control them. Nonetheless, neither chemical reductions nor chemical additions appear to have had any significant impact on producers:

We were concerned about how to control the Roundup Ready canola. It hasn't turned out to be too big of a problem. It has been fairly easy to control. . . . Chemicals that you use in wheat generally wipe out the canola pretty easily. (SK#3, seed dealer/GM producer)

Some producers were irritated when they found canola that could not be killed with Roundup volunteering in their yards and between their trees. As one producer complained, Monsanto's claims that their RR canola would not spread was "bullshit," and the spread of RR canola was a great nuisance because 2-4-D had to be used in the yard instead of Roundup when it volunteered. Other farmers also had to respond outside of their traditional use of Roundup, sometimes even removing the plants by hand. The problem does not seem to be overly significant yet, but raises flags in regard to future applications of the RR trait.

The most significant environmental issue raised by producers is that of weeds developing resistance to glyphosate, although it predominantly remained in the realm of a potential issue. Outside of the debate over RR wheat, where it was given a fair amount of play, and despite extensive coverage by environmentalists,[3] the issue appears to currently be of low concern to farmers. Canola is only one of a variety of crops grown in a farmers' rotation, however, although the higher price and the improved weed control have increased the number of acres grown. While there is little indication of resistance, as yet, the fact that farmers can switch to another oilseed crop, change to another herbicide-resistant canola, or increase their cereal or specialty crop acres also likely limits their concern over its development. Nonetheless, there is caution: for example, one farmer explained that while resistance was not yet an issue, he used LL canola so that he could still use Roundup for burn-off, while reducing the chance of resistance to it developing on his farm.

But Does It Help Make a Buck?

In short, while there are certainly those who disagree with particular aspects of the herbicide-tolerant agricultural biotechnologies, most users found their physical properties to be largely beneficial. It is possible the technology has been slightly too beneficial, as a number of respondents ruefully noted that the vast increase in Saskatchewan canola acreage has been followed by a reduction in prices. Others argue, however, that the price of canola is set by the price of soybean oil from the United States and other markets, and therefore canola volume has no impact on price. While it is not possible to consider economic returns unbounded from the issue of commodity price, it is possible to estimate the relative economic benefits of the technology in comparison to conventional canola. With respect to these economic benefits, the conclusion is far more ambiguous than for the technologies' physical attributes.

The economic ambiguity could have something to do with a given producer's prior practices. For those who previously could not grow canola, for example, the GM varieties provided a new and economically beneficial cropping option: this is in contrast with those who were already growing the crop and only changed inputs. In any case, while a number of producers claimed the technology gave them an unambiguous economic benefit due to gains from weed control, a significant number of others were appreciative of these gains but remained ambiguous about its economic ones:

> It's been convenient for us on the farm, that's for sure. It's allowed us to grow canola crops that have a lot fewer weeds in them. Yeah, weed control has gotten a lot easier. And I was going to say cheaper, but I'm not sure of that. You've got your technology fee of $15/acre and another $5 to $10 of chemicals. So overall I'm not sure it's cheaper, but it's more effective. (SK#3, seed dealer/GM producer)

Similarly:

> Since 1999 I'd say I was farther ahead [economically] with growing altered gene variety canola instead of open pollinated. Some years it's been close, there really hasn't been a lot of difference, but I would say since 1999 it's probably been a benefit to me. (SK#29, GM producer)

By all accounts, the cost of the RR and the LL herbicide systems was fairly similar. Roundup Ready canola was sold with one of the cheapest herbicides, Roundup, which had gone off patent and therefore faced competition from generic chemicals. However, RR technology was only sold with a $15-per-acre Technology Use Agreement [TUA]. As one grower explained it, Monsanto's seed plus the TUA together cost about $35 to $40 per acre with an added $5 to $6 for chemical application. The cost of Bayer CropScience's LL canola seed was similar to Monsanto's but had no accompanying technology fee. Bayer's Liberty chemical was still under patent, however, and was priced sufficiently high that the LL and RR technological packages ended up being on par with each other.

Producers of both kinds of genetic technologies often noted the increased cost of seed. Discussing drawbacks, one LL grower stated:

> Probably the initial drawback would be — and one that you notice the quickest — is the increased cost of seed. That would be the only drawback that I can think of. (SK#29)

While overall the cost of GM technologies was considered high, for many they weren't specifically singled out, but were just one of a number of high input costs a farmer vainly struggled to balance against low commodity returns. The voice of the farmer from Saskatchewan is clear: inputs are rising, commodity prices are low, and farmers are finding it harder and harder to survive. The end result is farmers pushed closer to the financial edge:

> Your seed costs and of course your sprays are going up every year. It's getting to be a bigger risk all the time. A guy used to be able to absorb one or two failures; now one failure can damn near sink you. Because everybody is maxed out on their operating loans, and their credits, and that kind of stuff. (SK#25, GM producer)

One drawback that was specific to the increased cost of transgenics is that it is an up-front cost, due before the crop — and any potential benefits — has been harvested. While the added input cost might provide a net economic benefit when the harvest is good, it increases the risk when drought, frost, or other factors negatively affect crop productivity. Further, rising prices were noticed by both producers and stakeholders. As characterized by an agricultural consultant:

> Growers increasingly are saying, "Gosh, there's something out of whack when canola seed costs continue to go up and up." You know, when a pound of canola seed is worth more than a bushel of canola that you harvest, there's something wrong. Fifty pounds to a bushel. (SK#22)

Talk of rising prices was often associated with a sentiment that producers couldn't do anything about it. The concern was such that producer organizations were beginning to discuss how to address the issue. For example, the Saskatchewan Canola Development Commission [SCDC], a grower organization funded by a check-off,[4] was considering entering into canola breeding in order to provide some means to address rising seed costs:

> We are evaluating should we get involved in breeding programs to pro-
> vide competition for industry. . . . So that's the evaluation process we
> are in. To try and figure out what is the best way to provide reasonable
> priced seed to growers. . . . It's one of the major issues we have in the
> organization right now. (SK#28, SCDC)

While producing conventional canola may have drawbacks, to be discussed, canola is only one of a typical Saskatchewan farmer's potential rotations. Those who find the cost—or the financial risk—too high could choose other crops. Potentially, this has a dual cost reduction: first, another crop would not have the premium charged on GM seed, and second, farmers would be able to save seed from nonpatented conventional crops. Strategy is thus essential in crop selection:

> I probably seed less [canola] in the last couple of years, because of the
> expense, yeah. . . . Like last year I put in 250 acres of mustard, because
> I had my own mustard seed here, and it worked just fine, and of course
> that's conventional as well. So instead of canola, I might put more flax in
> or something; flax is cheaper to seed too. (SK#26, GM producer)

The net benefit of such strategizing depends, of course, on the prices the alternative commodities fetch in comparison with the traditionally higher priced canola. Given the vast acreage seeded to canola each year, a majority of producers are clearly still finding it worthwhile to plant despite any economic ambiguity. This begs the question: why?

The answer is readily apparent after a few conversations with farmers. Unlike in Mississippi, as will be discussed, the introduction of agricultural biotechnologies has not had a make-or-break impact on the viability of canola. Saskatchewan farmers are market dependent, however, and do not have the extensive agricultural subsidies that can ease the financial burden of farmers in the United States. Further, they cannot pass any financial imbalances on to consumers, as they must take the price the market sets for their commodity. Consequently, the number one priority for farmers squeezed between high input costs and low commodity prices is yield, and herbicide-tolerant canola has allowed farmers to increase yields by decreasing weed competition and dockage, allowing them to farm previously fallow fields, and allowing them to farm more acres with the same amount of labor and equipment. Alternately, GM farmers can farm the same number of acres, but with less effort, which enables them to supplement their incomes in other ways. Unfortunately, any increase in canola yields gained through GM technologies was not so much characterized by benefit as by necessity. Two GM growers discuss how their yields have increased, without any improvement to their incomes:

> When I first started farming in 1978 I grew open pollinated varieties and my costs were .25/lb, and I sold that canola for $6 a bushel. That was back in 1978. Whereas now my costs are $6/lb and I'm still getting $6 a bushel. My yield has gone up, like before my yield was maybe 30 to 35 bushels an acre with conventional open pollinated varieties, whereas I have gone up to maybe 150 bushels an acre with Liberty Link canola, but still with the prices the way they are now. . . . Basically what we are doing with altered gene varieties is we are working on yield, we are not working on price, and that is why guys are growing it. (SK#29)

> You just keep getting bigger yields all the time, sell more, but you still get the same as you did 40 years ago. (SK#25)

The constant push is to get an edge on yields and stay ahead with that edge. While technology may not provide farmers with a better income (although for some it does), it can help them stay in the game. As one seed dealer and GM producer characterized the motivation for adoption:

> I think if farmers made more money, you know, in the end, I think they wouldn't have to turn to some of these things. It really becomes an economic issue, really. Strictly a survival issue. (SK#20)

Those who don't adopt, fall behind. In this sense, for many the technology is not about economic benefit, but about avoiding economic disadvantage—a "technological treadmill" scenario familiar to many industries (Cochrane, 1979). There were those who noted the temporality of such technologically induced gains, as prices would soon drop in response to increased yields; nonetheless there was always the hope that new technology would arrive to take them the next step to profitability.

In fact, the hope that biotechnological developments in canola will provide producers with new marketing opportunities and give them a global edge was pervasive in Saskatchewan. Simply marketed as a vegetable oil, canola is competitively disadvantaged against the cheap and abundant soybean oil, particularly given transportation cost differences and increasing competition from locations such as Brazil; there is only so much farmers can do with efficiency. Specialty types of canola, such as high-erucic-acid canola (for industrial uses) and high-oleic canola (a healthier, more stable oil for health-conscious consumers), are marketed for a premium, however. There is thus a great deal of interest in expanding these specialty oil markets, with canola considered particularly suitable for engineering with specialty traits. Therefore, in addition to biotechnologies' potential for yield increases, they are touted as the harbingers of niche marketing opportunities. This hope for the future of GM canola is held not only by industry and stakeholder organizations, but also by those in the farming community:

> The trends are costs will continue to escalate. Market prices won't keep pace with it, not likely, and that squeeze will continue. The margin between your costs of inputs to what you get in the marketplace is going to continue to narrow. There are a few options being tried; one involves the biotech industry, and that is to engineer a type of crop that can be grown in an area where you haven't got foreign competition or competition in another region. . . . somehow biotechnology [will] take us into an arena that will allow us to produce in this region, and for a few years have an opportunity because you are not coming up against competition with that particular crop in a foreign country. (SK#11, GM producer)

Who's in Control?

Monsanto's Technology Use Agreement

The costs and benefits of using GM crops are not restricted to their agronomic and economic impacts, however. A significant question raised by agricultural biotechnology concerns the issue of control: will the use of biotechnologies reduce a producer's control over his or her agricultural production? A significant target for those who have these concerns has been Monsanto's Technology Use Agreement [TUA].

For producers, the ethics of Monsanto's TUA appeared nearly as irritating as the cost itself. Many felt that, given that the justification for the TUA was the need for research and development dollars, the money had long since been recouped:

> Yeah, [the TUA] bothered me right from the beginning. I didn't believe in that $15. Then after a while, OK, well they have to have it. But that was years ago. They don't need that no more. That's all been paid a million times over. If that's what they were using it for, their research, they haven't done any more, so quit it. (SK#25, GM producer)

Given the proximity in cost of the RR and LL systems when you calculate their total technology package, the irritation over Monsanto's TUA went deeper than strictly with respect to its cost. While economically and agronomically the two technologies have a great deal of similarities, negative expressions are largely reserved for Monsanto. The TUA and its conditions of use are where one of the more significant differences between the technologies emerges. As noted, Bayer switched to hybrid production of its GM canola fairly early on. With the quality limitations of second-generation seeds from hybrids, and with the price premium for their patented chemical, Bayer was relatively assured of capturing a return on their investment without any further contract provisions. Monsanto, on the other hand, could not extract such returns from its off-patent herbicide. Further, its herbicide-tolerant trait was instilled in open pollinated varieties; thus there was no physical disincentive to using second-generation seed the following year. Consequently, Monsanto used the TUA, which included a $15/acre fee and a number of provisions that a grower had to follow in order to use the seed.

As a number of stakeholders characterized it, producers objected to Monsanto's TUA because psychologically they cannot see any benefit to what they are paying for. For example:

> You always hear complaints, I mean, nobody likes to pay something for a piece of paper, right? I mean, you'd like to see that in your pocket instead, right? But it's the format chosen by Monsanto to capture profit and the return on their investment. (SK#17, Saskatchewan Wheat Pool [SWP])

Similarly, a representative from the Canola Council of Canada [CCC] states:

The biggest thing that I have seen with TUAs is that—and I'll speak from a farmer's perspective here—I don't like to be controlled by anybody. If Monsanto wants to control me, they can kiss my butt. And yes I did sign it, but in my heart I don't like to be controlled in that fashion. . . . The idea of paying for technology you are going to use is not a bad idea, but Bayer CropScience does it, and they don't charge you TUAs, they charge you every time you use the product. (SK#4)

As noted, however, it is not strictly the "piece of paper" that offends farmers, but what is written on it as well. Monsanto's grower contracts have a number of provisions—a prohibition on seed saving, the obligation to sell the end-product to a Monsanto-approved processor, and the right of Monsanto to inspect a grower's fields for three years—that a producer must contractually accept in order to purchase the technology. These provisions can elicit some strong responses from producers:

I don't like Monsanto; I feel they are heavy-handed. They want to control the product from the time it goes in the ground to the time it goes into the consumer's mouth, and you're just a pawn, you know, their servant really. I don't use Monsanto products. I don't really care for their attitude. (SK#19, conventional producer)

When you sign the agreement, you know, you pretty well sign all your privacy away; they can come onto your land and check your farm three years from now, and check your bins and do all those things, and when that first came out it really rubbed me the wrong way. I walked out of the first meeting I went to. I just said there's no way I would sign up for this sort of—I don't know if invasion of privacy is the right word—but eventually I caved in. I got mad the first meeting and told them so, and told them what I thought. There's quite a few guys that when it first came out were rubbed the wrong way by that. It's not just about the $15 per acre, it's about what that $15 represented. (SK#14, GM producer)

As the latter producer described it further, he eventually saw the benefits of the technology demonstrated by others and finally accepted Monsanto's conditions, which, in practice, he did not consider as fearsome as they initially seemed.

Ultimately, most producers I spoke to did not feel the company had abused its contract agreements or that it was in any way nefariously out to get them. Most simply concluded that if they didn't want to sign the contract, they could just decline using the technology. The issue of liability remains, and will be discussed in relation to the Schmeiser case. With respect to a social reorganization of agriculture, however, the restrictions on seed saving need further consideration. While accepting such restrictions with the use of biotech crops has been considered a worthwhile trade-off for many producers to date, in the context of rising costs and the prospect of an increasing number of commodities marketed under this system, such restrictions take on a different meaning.

The Rise of Biotech and the End of Farm-Saved Seed?

Prior to the introduction of biotechnology, canola seed was relatively inexpensive. In addition, farmers had the option of saving seed for use the following year, which included the additional cost of treating the seed. For the majority of farmers, this was considered too much hassle given the small economic benefit. Consequently, there wasn't a great deal of canola seed saving even prior to the introduction of GM technologies, although it was certainly practiced by some. For those who did save seed, the use of patented biotechnologies represents a full shift from previous farm practices:

> When I was growing conventional seed prior to 1999, I saved seed and got it cleaned and treated. I probably did that for maybe half a dozen years. But since I've been growing altered gene varieties like Liberty Link I've never kept a seed or cleaned it or anything like that. I've been buying certified seed every year. (SK#29, GM producer)

Even among those who saved seed in the past, most expressed no strong negative feelings about the practical restrictions on seed saving. At its simplest, the trade-off was a deal they were willing to take.

Hypothetically speaking, however, many producers believed that seed saving was an important right to maintain. The issue of farmers' right to save seed had even more salience among the better informed in the community, given that the federal government had funded a Seed Sector Review to consider revisions to seed regulations. A serious concern over the review was that it would prompt efforts to revoke the farmer's exemption in the Plant Breeders' Rights Act, and would require producers to purchase new certified seed every year. In the eyes of many, the right to save seed was under significant threat. While farm-saved seed was already negligible in canola, this was not the case for other crops. Wheat, for example, is considered more stable in retaining its characteristics, and only about 5% to 20% of wheat seed was purchased anew each year (depending on whom you ask). Thus restrictions on farm-saved seed in crops such as wheat would be significant.

However, the producers interviewed found the trade-off of seed-saving restrictions for the technological benefits of GM canola worthwhile. Further, while there was support for farm-saved seed, it was balanced against the importance of new technologies for Saskatchewan farmers: if technology developers did not gain sufficient returns on their investment, new technologies would not be forthcoming. The balance of this perspective is particularly well noted by those in the contradictory position of seed dealer and farmer:

> You know, all of us farmers would like to save our own seed, but if the companies that develop the varieties can't see their way to a profit then they are likely to pull out their investment, and maybe not give us improved varieties as we go along. So it's kind of a dilemma there. (SK#3, seed dealer/GM producer)

> Even as a seed grower, I'm in favor of farm-saved seed, you know. We've got a real dilemma in our seed industry, you know, because farmers aren't making a lot of money, so they're really in a cost-price squeeze. . . . So if farmers are in a tight squeeze they don't buy certified seed. They will back up to the bin, or the portable cleaner will come in to clean it, or they will clean it themselves, and put it back into the ground again. You know, so then you see . . . like the patents . . . they want to protect what they developed. I mean, it costs a lot of money to keep people around to do those things, and I don't blame that, but are they really making more and more money all the time? Do they have to make as much money as they are? (SK#20, seed dealer/GM producer)

While regular seed saving may not have been an important part of canola production, the option to do so does provide flexibility during times of need. Using farm-saved seed is one economic strategy when farmers have low funds, for example, such as in the year following a bad crop. Another economic strategy is to alternate farm-saved seed with certified seed, intermittently purchasing certified seed in order to upgrade to better varieties. The importance of such flexibility became particularly salient in Saskatchewan when successive drought and frost had significant economic impacts. In August of 2005, after two years of drought, Saskatchewan suffered a killing frost that caused approximately a billion dollars in losses. These environmental curves compromised the economic viability of many Saskatchewan farmers. Indicators of the trouble could be seen in auction sales and loan application declines:

> You know there are places in Saskatchewan where a third to a half of farmers that are going to the bank to arrange their operating loans this year get turned down. (SK#2, academic, crop development)

In addition to the issue of bad years, the economics of buying seed has changed. Historically, the economic motivation for saving canola seed was low, but with the almost tripling in seed cost since GM technologies were introduced, the economic incentive for saving seed has greatly increased. While Roundup Ready seed is prohibited from being saved by contract, saving Liberty Link seed (while also patented) has only been prohibited in practice through the yield reduction that comes from growing a second generation of a hybrid. Unlike corn, however, where this yield reduction prohibits a viable second-generation crop, the yield losses from saved hybrid canola are not that dramatic. In tough times, absorbing the yield loss might be the best economic strategy, and by 2005 there was increasing evidence of this strategy emerging:

> In Saskatchewan we had two years of drought, one year of a bad crop, and another year was a mediocre crop, and economic conditions—a lot of farmers were saying, "Hey, I can't afford to do all this stuff. I can't afford to take the risk." And a lot of farmers were [saving hybrid seed]. (SK#28, Saskatchewan Canola Development Commission)

> With people pushed for resources and with seed costs going up, which is a big issue that I think is going to get more and more important, there have been people who are holding back and reusing hybrid seed. (SK#22, agricultural consultant/media)

While hard statistics are impossible to obtain, numerous stakeholders estimated that the amount of farm-saved hybrid seed was increasing. A respondent from the Saskatchewan Wheat Pool, for example, estimated a 10% to 15% drop in the amount of purchased certified seed:

In canola it's been as much as 95% purchased every year. I mean, now it's fallen 10% to 15% lower than that, because of the high seed prices and some of those growers not seeing benefit for those purchases, and they are trying to look at alternatives, in other words, keeping some of their own. (SK#17)

Given the rise in seed saving, in 2004 the Saskatchewan Canola Development Commission initiated research into the yield loss farmers could expect when using saved hybrid seed. During data collection, it became evident that hybridization was a barrier to seed saving only if farmers were sufficiently economically stable. When economically stressed, many felt pressured to reuse seed despite any yield reductions. Unfortunately for producers, Bayer addressed this strategy by following Monsanto's lead. As of January 1, 2005, Bayer put new labels on their seed bags outlining their "conditions of sale." While Bayer did not introduce contracts with extra provisions, as did Monsanto, these seed bag labels explicitly affirm their patent and their right to prohibit the saving of their seeds, which "shall be used only for planting of a single commercial crop" (copy with author).

In addition to the loss of an economic strategy, the prohibition on seed saving has had some unexpected impacts. In 2005, for example, Manitoba experienced extremely heavy flooding, making fields too wet for many farmers to seed their land. Consequently, these farmers only had volunteer crops from the previous year, including crops of GM canola. In the past, a volunteer crop would be better than no crop, but now patents prohibited growers from using the second-generation GM canola. The incident offered an important glimpse into unexpected losses that farmers can face when compelled to accept seed-saving prohibitions in order to access GM technologies. In response to the disaster, Monsanto affirmed its TUA. Bayer CropScience made a special regional exemption, however, allowing Manitoba producers to cultivate the crop where they could not plant another.[5] A less well-publicized and widespread problem in the future may not result in the same exemption, particularly once the technologies have gained greater social acceptance.

The Alternatives

While many find that the benefits of biotechnologies currently outweigh the drawbacks, the rising seed costs and periodic economic stresses indicate that this might not always be the case. An important control issue, then, would be whether producers have viable alternatives in the event the cost-benefit balance tips against biotechnologies' favor.

Strictly with respect to availability, conventional canola seeds are still an alternative—in the specialty varieties produced under contract and in the shorter germination *Brassica rapa*, but even in small amounts of open pollinated *Brassica napa*, as firmly asserted by a representative of the Canola Council of Canada:

> You still have options. You don't have to buy their seed. You can buy other commercially available seed. . . . They are available to you. You aren't restricted. That's a media hype, in my mind. (SK#4)

Somewhat less adamantly, one producer responds:

> Yeah, it's still there. (SK#29, GM producer)

The response sums up both the availability and the enthusiasm most growers have over any conventional canola that has no compensatory premium attached to it, as do the specialty canolas. While technically available, most producers had no doubt that the GM varieties of canola were superior, not strictly because of their GM traits, but because varietal development in yield, disease resistance, and other non-GM traits have only been commercialized in association with private investment into GM herbicide tolerance. Therefore, while conventional varieties are available, they are not considered agronomically on par with the GM varieties. As frankly laid out to me by this and other growers, what little public varieties are available don't measure up in yield, so using them is a big disadvantage:

> As far as not growing any conventional varieties, I guess it would depend on . . . it all boils down to yield in the end, and what you are getting for your money. (SK#29, GM producer)

Similarly:

> No, everything I've experienced [in conventional production] is a disadvantage because it doesn't yield as well. And you need every bushel in order to keep going. So that's the main reason why guys are changing, is yields. (SK#25, GM producer)

As private industry has taken over the development of new canola varieties, their emphasis has been on GM canola. Thus while herbicide tolerance is the only GM trait, the packaging of this trait with other varietal improvements has made the GM seed package superior to conventional ones. Therefore, the strict availability of conventional canola notwithstanding, complaints over the atrophy of public canola breeding and the resulting decline in the competitiveness of conventional seed are fairly common:

> Some of the conventional varieties competed pretty well, but as time moves forward and nobody does any work on them, they have lower yields, they have less disease resistance. (SK#22, agricultural consultant/media)

In short, the development of conventional canolas has fallen by the wayside. Should farmers want to return to such varieties at some time, they would not be able to garner the yields that have become the norm for making a return, among other drawbacks. The agronomic disincentive to reject GM canola therefore becomes greater as time goes on.

But why should canola development be dominated by private industry? According to some, the cost of obtaining regulatory approval is so high that only major companies can go through with varietal registration, and this accounts for the decline in public breeding. According to others, the idea of "biotechnology" took on such a life of its own that it overtook all other breeding—a case of the tail wagging the dog:

> [Biotechnology] has kind of taken over from traditional plant breeding, and it is actually now the focus from which plant breeding is managed. Instead of a tool, the tool is running plant breeding, instead of the breeding tool being looked at as just a tool, and if it is economically wise to use it or not. (SK#17, Saskatchewan Wheat Pool)

The decline of public breeding is also consistent with a general trend of government cutbacks of traditional services in agriculture. In 2005, for example, Saskatchewan-area agronomists were discontinued, and with them went hands-on consultation and personal assistance to producers, which is now available only through telephone service to the province's Agricultural Knowledge Centre.

Whatever the differential weight of these factors, it is clear that while the government previously supported extensive research into varietal development through breeding, its focus firmly shifted to biotechnology and to supporting industry rather than commercializing varieties itself. The level of public versus private breeding depends to a large extent on the crop at issue. While canola has gone to private industry, public breeding still occurs in crops like wheat and barley, and grower organizations are active in other commodities, like pulses.

Whether the progressive exiting of public breeding from varietal development is a retrenchment or a redefinition of purpose strongly depends on one's position in agriculture, however. Some, like Ag-West Bio, Inc., a nonprofit organization created to facilitate and promote the bioeconomy in the province, think that biotechnology represents the evolutionary climax of plant breeding:

> So if you look at all the major crops that are grown on the prairies,
> they are all at a different stage of evolution. So if you take the wheat
> industry as an example, almost all of the varieties on the market in
> the wheat industry come from public breeding programs. Why? Be-
> cause private breeding companies are not engaged with wheat breeding
> to a large extent . . . And why aren't they involved? Because they haven't
> found a value capture mechanism that would give them a comparative
> advantage in the marketplace. So in that sense the wheat industry,
> I would say, is less developed [than the canola industry] because the
> public purse has to pay for the development of varieties and it is not in
> the private sector where it should be, in my view, because in the pri-
> vate sector you are more likely to get newer things happening faster, and
> that will eventually be of more benefit to the producer as long as there is
> choice and competition in a free-market system. (SK#27A, Ag-West Bio,
> Inc.)

Under this new orientation of public research, public organizations like the Plant Breeding Institute, Agriculture and AgriFood Canada, the Saskatoon Research Centre, and the like, do the background work on desirable traits in canola that the private industry picks up and brings to commercialization. One drawback of this partnership is that when industry does not predict a trait to be suitably profitable, it is not developed, no matter how beneficial it might be for farmers. This division of labor is consequently perceived negatively by some, such as the Saskatchewan Canola Development Commission:

All these [public research facilities] have developed traits that have potential to provide benefit. But the only way you get that trait into the marketplace [is this]: most of those people now are developing germplasm or identifying genes and marker genes, so they'll come up saying, "Here's the germplasm, we've identified this gene, here is how you find it, this is how you transfer it, and it has drought tolerance." Then they give it to the company or they put it on the market saying, "What would a company give us for this trait?" Then the company comes in and says, "Well, I don't know, I'd have to get 50% return, if I have to put it in our varieties, there's all the development work plus there's all the regulations, I have to cover all those costs, I don't think it's worthwhile." Where does it go? It might be a benefit to our growers, but we can't get it through the process. Now [canola's] the only crop where that's happening. . . . Canola is actually the only one that has been taken mainly over by the seed companies. (SK#28, SCDC)

The retrenchment of the public breeding programs is therefore a significant concern to grower groups, who are left dependent on the decisions of private industry. While public breeding is conducted for the benefit of producers, private industry commercializes traits based on profitability, which may not necessarily coincide with what is best for producers. In some cases, important traits that would greatly benefit producers have already been identified, but lack the industry will to bring them to commercialization:

And that starts to be a concern for growers to say we have all these traits that we know are there. These are realistic traits, they are already identified genes for drought tolerance—and you can imagine what kind of impact some kind of drought tolerance could have on canola in Saskatchewan. (SK#28, SCDC)

Particular sectors, regions, or groups can also be left out as private industry steers the course of plant breeding toward whatever is most profitable. For example, when it appeared that the returns on *Brassica rapa* (the canola variety appropriate for the north) would not be sufficiently significant, industry pulled out of all *rapa* breeding to focus on *Brassica napa*. Without public breeding, those northern varieties would languish. Ultimately, the Saskatchewan Canola Development Commission and the Alberta Canola Producers, in conjunction with Agriculture Canada, took over the *rapa* breeding program. Problems such as these, in conjunction with the rising cost of seed, have prompted the SCDC to consider the major investment of entering into canola breeding more broadly.

Conclusion

While the two types of GM canola—RR and LL—each has its own particular agronomic strengths and weaknesses, their physical properties have had a fair amount of commonality of impact. The introduction of GM canolas has assisted producers' ability to grow the crop (to the point of providing a new cropping option for some), reduced losses due to weeds, and provided a handful of ancillary benefits, such as reducing the manual labor required, allowing producers to work off-farm or expand their acres, following the economic strategy of volume production. The use of the technology can also increase management drawbacks, but overall it appears to provide producers with more agronomic choices. Economically, the impact is more ambiguous: biotechnologies can provide economic benefits, particularly for new adopters, but this benefit can be lost when the harvest is poor or commodity prices are low. Further, there is risk involved in higher input costs, particularly in frost- and drought-susceptible Saskatchewan.

The use of agricultural biotechnologies cannot be considered in strictly agronomic or economic terms, however, as they are associated with a package of shifts in agricultural practices, some whose significance will not be revealed for many years. This book is concerned with the question of control, and the extent to which these technologies are facilitating a reorganization of agricultural production that shifts power and control away from farmers. A number of sites suggesting such a reorganization were apparent here: restrictions on seed saving, liability issues, Monsanto's TUA, and the decline in public breeding. The Schmeiser case notwithstanding, the issue of liability for inadvertent infringement did not appear to be a significant concern in Saskatchewan, and the potential for it was reduced by the high GM adoption and low level of seed saving already practiced in canola production. Another GM crop introduction that was less readily adopted could increase the importance of the issue, of course, and provide further insight into the evolving legal relationship between farmers and biotechnology developers.

Monsanto's TUA was an issue unto itself, and many farmers visibly bristled at being told how to conduct their affairs. Nonetheless, the benefits of the technology frequently outweighed this objection. The reason why is significant. Saskatchewan farmers survive in the margin between low commodity prices and high input costs. Farmers pushed to very small margins are motivated to take on any technology that might assist in their survival, and biotechnologies can play a short- and long-term role in this: in the short term, small increases in yields (or reduced dockage) expanded over enough acres can make the difference between debt and profit; in the long term, biotechnologies might provide new market niches. Saskatchewan farmers' market dependence consequently requires some delicate balancing between these strategies and the need to avoid upsetting their customers, notably those in the European Union.

From a micro perspective, as long as a producer can opt out, agricultural biotechnologies are simply another production choice. As we saw here, though, the decline in public breeding makes conventional canola increasingly unattractive. In the context of an agricultural reorganization, this decline in public breeding is highly significant in itself. Accepting the terms on which biotechnologies are offered further costs farmers their traditional seed-saving rights, and denies them an important economic strategy. When commodity prices are high and harvests are good, the impact of this restriction may be negligible. During hard times, however, the postulation of a mutually beneficial relationship between agribusiness and farmers breaks down (Pfeffer, 1992). Conflicts of interest then start to become apparent, as evidenced by Bayer curtailing the already desperate tactic of resaving hybrid seeds. Not to overstate the case, restrictions on seed saving are not the only nail in the economically imperiled farmer's coffin; a reasonable assumption, however, is that the reduction of economic strategies forces those who are already economically stressed out of the industry that much quicker.

As long as producers can rotate to different crops, they still have alternatives to biotechnologies, but as more GM crops are commercialized, these alternatives are increasingly closed off. In this context, the issues that arose around RR wheat and the Schmeiser case—discussed in the following chapter—have particular salience.

CHAPTER 4

. . . And the Fall of Wheat

There would have been a feeling that Monsanto's trying to push this regardless of what people want, or what's the right thing to do, or what's best for everybody. Right, it was good for them, but was it good for the people they were supposed to be trying to help?
SK#18A, SASKATCHEWAN ASSOCIATION OF RURAL MUNICIPALITIES

It could affect us. Our neighbor just owns it and it blows over to us the way Percy claims it happened to him, which I strongly believe that's what happened—yeah, organics could be a thing of the past. Definitely.
SK#12, ORGANIC PRODUCER

Introduction

While resistance to biotechnologies from local Saskatchewan environmental groups is low, it has pushed forth in other avenues. In addition to the resistance raised by those supportive of Schmeiser, biotechnologies have garnered sector-specific resistance from organic producers (opposed to GM crops in general) and broad-based but temporal-specific resistance to Roundup Ready wheat (primarily on the basis of market acceptance). Wheat production in Saskatchewan has dropped significantly (over 55% between 1996 and 2001 [Saskatchewan Agriculture, Food and Rural Revitalization (SAFRR, 2002b)]), as low wheat prices have motivated some producers to switch more toward livestock and forage crops. Nonetheless, wheat remains the number one crop in the province. At the same time, as already noted, Saskatchewan is a key locale for Canada's burgeoning organic agriculture industry. Thus resistance from these two sectors has not been a minor inconvenience, but has significantly informed the context of biotechnology development in Saskatchewan.

Opposition: Greens, Lawsuits, and Wheat

Clearly those who choose to use Monsanto and Bayer's GM canola find it beneficial. This does not mean it is complaint free, as is most particularly evidenced with respect to Monsanto's TUA; but while a few farmers had strong feelings about Monsanto, they did not necessarily feel sufficiently negative to switch to Bayer's product because of it. Nonetheless, there has been opposition to biotechnologies in the province. In fact, there has been sufficient opposition to render many industry supporters wary and beleaguered, and sometimes downright testy. During a scheduled interview with Ag-West Bio, Inc., for example, interviewees required that every question's terms be defined to the point where asking anything at all became nearly impossible. Even those seemingly benign queries that survived this process prompted off-topic (but "on-message") packaged responses—this despite assured prepublication screening of any quotes. Perhaps in partial explanation of this excessive wariness, one of the interviewees offered this:

> There's a lot of recycled arguments and misinformation that exists in the whole spectrum, and some really vested interests and activists, and it's a very politically, socially sensitive and active area. (SK#27B, Ag-West Bio, Inc.)

While this was an extreme case of subject sensitivity, it was certainly not the only one, and a handful of other stakeholders also requested the right to prescreen quotes. No one with an interest in the well-being of the industry appeared confident that even an academic interview would not become immediate media fodder, complete with out-of-context damaging quotes. Yet, despite all the testiness, there was really no anti-biotech activity from any local environmental organization. Most likely, given the economic pressures on the agricultural community, further threatening agricultural livelihoods would not be beneficial to the long-term survival of any such organization. In any case, the only public anti-biotechnology sentiments from a Saskatchewan organization came from the organic industry. This apparent environmental free ride raises the question: how had the province's biotechnology industry and promoters become so testy? What was the threat to their industry?

In the national context, environmental organizations such as Greenpeace, the Council of Canadians, and Friends of the Earth have taken biotechnologies on as an environmental issue. While these organizations already have an impact on their own, they further their effectiveness by working with and publicizing local issues, such as the recent attempt to declare Prince Edward Island a GMO-free zone. In Saskatchewan, the Schmeiser and the Hoffman litigation and the attempted introduction of RR wheat were high-profile issues that garnered exposure and consequent widespread public debate. This publicity raised a significant amount of anti-biotech support and stole a fair bit of wind from the Saskatchewan bio-economy's sails. I will discuss each of these local issues in turn, beginning with RR wheat.

Take Our Wheat!

In early 2000, Monsanto initiated the long process of gaining regulatory approval to commercialize its RR wheat in both Canada and the United States. Unfortunately, it neglected one of the golden rules of business: provide a product that your customers want. Opposition bubbled on both sides of the border, particularly from wheat growers concerned about the potential loss of markets. Post-Starlink, concerns about markets were paramount. Even leaving aside the issue of whether farmers could sell RR wheat, many were not keen on the product for strictly agronomic reasons. Unlike canola, wheat is not as vulnerable to weed pressure, and there are fairly effective and reasonably priced chemicals available for what weed control is needed. Consequently, even the most benevolent of responses to RR wheat expressed a fair amount of disinterest in the product:

> There are enough chemicals out there for wheat now to keep a nice clean field; I don't know why you'd want a field of Roundup Ready wheat. (SK#26, GM producer)

Wheat is also a lower-value crop, and increased input costs—as with the introduction of a TUA—would push the margins of return very low, even if the TUA fee were reduced below that of canola, as suggested. Seed saving is also prevalent in wheat, and restricting it would represent another significant cost increase for wheat producers. In view of these factors, RR wheat would likely cost farmers far more than conventional wheat—without providing the agronomic benefits associated with herbicide-tolerant canola. Given the economics alone, interest in RR wheat was already lukewarm. In a context where RR canola was already prevalent, serious agronomic issues were also raised by the prospective extensive use of Roundup.

Glyphosate, the chemical in Roundup, plays a significant role in modern farming practices as a chemical treatment to "burn" the fields in the spring, prior to planting. In the practice of crop rotations, volunteers from the previous years' crop become weeds in the next. While volunteer RR canola complicated farm management somewhat, it was manageable through an additional chemical additive to the glyphosate. The addition of RR wheat to the crop rotation would create further complications (as wheat volunteers in canola would be resistant to Roundup and vice versa), would require more chemical additions at increased cost, and would increase the chance of weed resistance developing to a highly prized chemical. The contradiction inherent in promoting RR wheat as a new tool is made clear by a representative of the Saskatchewan Wheat Pool:

> Why do you want to incorporate another tool that loses you other tools on your farm? Roundup is already used so much already, by incorporating its use as was proposed on wheat, it would mean that you lose its effect and benefits on other crops on your farm. So Roundup Ready canola wouldn't be a good thing anymore, and Roundup Ready burn-off of crops and stuff, because now these all become new weeds in your crops. So how many places do you want this technology? (SK#17)

Such concerns were repeated by many:

> I mean this spraying Roundup in the spring for burn-off, then you are going to spray Roundup in-crop spraying, then you are going to spray Roundup in-season, then you are going to spray Roundup next fall for burn-off—you are going to create a monster here. . . . You are going to create some weeds you are not going to get rid of, and then they'll be breeding with other weeds—oh the hell you'll come up with pretty soon. (SK#25, GM producer)

> Most people here don't till their land before they seed it to kill weeds; they spray it with Roundup either just before they seed it or just after they seed it. So do you want to use even more Roundup and maybe has-

ten the possibility of glyphosate-resistant weeds? Because if we didn't have glyphosate as a tool, it would really change how we farm. (SK#22, agricultural consultant/media)

Both economically and agronomically, then, there was very little interest in RR wheat. In the perception of many, it was useless at best, and most likely detrimental. This was not the total of objections to it, however. Hypothetically, RR wheat could have been commercialized without immediate negative impact on non-adopters (albeit the future risk of glyphosate resistance remained) if not for the issue of international marketability. Market rejection was a huge concern, however. As a number of producers queried: what was the need for a high-cost technology to increase production of a product that some—such as the EU—wouldn't even want? The general sentiment was that as wheat production was already sufficient to saturate demand and keep prices low, there was no reason to compromise demand by growing something potentially offensive to customers.

Given the impossibility of segregating RR and non-RR wheat in the current wheat marketing system, even a boycott of RR wheat by an overwhelming majority of producers would not be sufficient to preserve international markets. As long as even a few grew the RR variety, shipments would be contaminated and the probability of market rejection was nearly assured. Certainly, an important factor in the viability of segregation is the question of tolerance levels, which many countries still lack. Nonetheless, segregation is generally considered unviable, and even limited RR wheat adoption in the country would mean that all Canadian wheat would be considered GM.

This marketing issue was not theoretical to wheat farmers, many of whom were also canola growers and had been keenly aware of the rejection of their canola by European markets. While this had been a small and calculated loss by the canola industry, the Canadian Wheat Board [CWB] surveyed its customers and found that over 80% would refuse GM wheat. Further, while commercialization of RR wheat was planned for a number of countries, Canada was furthest along in the process, and would be first in line for these market repercussions (SK#24, CWB). This was not a cross that the Canadian wheat industry thought it should bear:

From a safety perspective I don't think there's any concern. But from the perception of our customers that don't want to see that kind of contamination, we have to be ultra-conservative, because you know, the wheat board is telling us what to handle, what to buy, and what to deliver, and to what customer, and if the customer doesn't want that in there, then that forms our opinion. . . . You have to meet customer demand, period. (SK#17, Saskatchewan Wheat Pool)

Producers and producer organizations were thus reasonably concerned that RR wheat would cause economic devastation. Monsanto's push to introduce the crop appeared to be unimpeded by these concerns, however, which they continued to downplay. In response, concerned organizations began to publicize strong sentiments regarding the harm that RR wheat could cause the Canadian wheat industry. The rising opposition to RR wheat, unlike public sentiment over other agricultural issues, was widespread and uncharacteristically united. Canadian Wheat Board lobbying was supported by a large coalition of agricultural interests such as the Canadian Federation of Agriculture [CFA], the Saskatchewan Association of Rural Municipalities [SARM], the National Farmers Union [NFU], the Agricultural Producers Association of Saskatchewan [APAS], wheat organizations from Alberta and Ontario, and many others. Advertisements were published in the names of key agricultural organizations publicly asking Monsanto to withdraw its application and stating that Monsanto's continued refusal could devastate Canadian farmers. Organic producers, concerned about GM contamination, filed an injunction against the introduction of RR wheat in conjunction with their class action lawsuit over GM canola contamination.

Concerns that Monsanto would proceed despite the opposition were such that the CWB ultimately formed an alliance with environmental and social organizations, such as Greenpeace and the Council of Canadians, in an effort to further strengthen their lobby effort (Wells and Penfound, 2003). Some considered this to be an unholy alliance, however temporary, given that these organizations were against biotechnologies outright while the agricultural organizations were only opposed to pending market acceptance. The alliance itself is nonetheless testament to the vulnerability these agricultural organizations felt in the face of Monsanto's perceived disregard for their economic well-being.

In short, farmers and farm organizations were frustrated by the nonchalance with which their needs could be rendered irrelevant by Monsanto's motivation to maximize its profits. In the face of overwhelming opposition, Monsanto finally withdrew its application for registration in

2004—but not before serious damage had been done to the faith that it would conduct itself in the best interest of its customers. The sentiment expressed by many agricultural organization representatives regarding RR wheat was that Monsanto was nothing short of a bully, willing to force its agenda to the detriment of even its own customers:

> The industry feeling is that the Roundup Ready wheat issue was handled very poorly by Monsanto, because it was basically, just, "Here's the wheat, take it!" Just shove it down people's throats basically, is what they were doing. (SK#18A, SARM)

> They were wrong in not consulting more and not working more with us. They should have done that. They tried to heavy hand it through pretty good there for a couple of years. We had to get quite a pretty good lobby strength together. (SK#30A, APAS)

The repercussions of the RR wheat fiasco were significant. From the industry side, Monsanto doubtlessly lost considerable investment dollars. Further, the company's image took yet another beating, as there was little doubt for whose benefit the push to promote RR wheat was intended. As the representative from SARM further reflected:

> I think they lost some business over it. I think there was some bad will over it. (SK#18A, SARM)

This "bad will" was expressed not only by organizations, but also by farmers. While the majority of farmers constrain their production decisions to primarily financial considerations, where there was leeway, a few found ways to express their resentment:

> In fact, I don't buy any Monsanto stuff anymore. Roundup Ready canola is the only thing I've bought from Monsanto for years because of the Roundup Ready wheat. . . . They just kept pushing it. They didn't give a damn what anybody said, just as long as it lined their pockets is all they cared about. I haven't bought a gallon of Roundup in years. (SK#25, GM producer)

> There's always other products available. I just avoid Monsanto's products, that's all. . . . The GM wheat issue definitely has a bearing on it. (SK#19, conventional producer)

Even more significantly, the attempted introduction of a product so unwanted by producers highlighted their vulnerability to the profit strategies of companies such as Monsanto. Those who determined that RR wheat was a significant threat to their industry and livelihoods found they had no mechanism to protect themselves outside of their massive lobby effort.

Roundup Ready wheat was not the first GM crop to fall to lobby efforts. In 2001, a GM flaxseed developed by the Crop Development Centre at the University of Saskatchewan fell under similar marketing concerns. The European market, which was the destination of 60% of Canadian flax, was opposed to GM flax. With flax farmers fearing for the marketability of their product, organizations such as the Flax Council of Canada and the Saskatchewan Flax Development Commission pushed for deregistration (Warick, 2001). While not yet commercialized, the GM flax already had federal approval and was in the process of being multiplied up (stocks increased) by contracted farmers in preparation for commercialization. Instead, as a result of the lobby effort, the existing crops were destroyed and the crop was deregistered. As the director of the Crop Development Centre put it: "It would have been irresponsible of us to fight to keep it on the market, and face the possibility of our farmers losing an export opportunity" (quoted in ibid.).

In regard to RR wheat, Monsanto clearly did not share this concern for marketability. Nonetheless, given the unrelenting opposition, Monsanto would have been foolhardy to persist in the case of RR wheat, possibly even forcing the government's hand, with long-term regulatory implications. Despite its belated acquiescence, however, the huge imbalance of power that was revealed during the altercation led a number of organizations to lobby for regulatory changes. In essence, these organizations wanted market considerations to become a factor in regulatory approval. While RR wheat was stopped, producers remained vulnerable to similar unwanted introductions in the future, and producer organizations such as the Canadian Wheat Board wanted a more tangible means of protecting themselves than lobby strength and public support:

> Eighty percent of our customers said if you have genetically modified wheat, they won't buy it from Canada. However, the genetically modified wheat had the possibility of going through all our system, and getting regulatory approval from the government. . . . if it had met environmental safety, feed safety, and food safety. . . . We wanted consumer acceptance to be an issue when it came to regulatory approval for new varieties. (SK#24, CWB)

As more formally stated on the CWB website: "To ensure that the interests of farmers and customers are fully considered, the CWB's position is that a cost-benefit analysis should be conducted as part of the regulatory process" (CWB, 2005). The statement includes a number of conditions that need to be met for regulatory approval, including "widespread market acceptance" and "a positive cost-benefit throughout the wheat value chain with particular emphasis on farmer income" (ibid.).

While the Canadian Wheat Board's position is limited to wheat and barley, other organizations, such as the Saskatchewan Association of Rural Municipalities, are not similarly restricted. According to their position:

> We were proposing a regulatory solution. Whereas some organizations and the chemical companies were saying it should be voluntary: it should be incumbent on the company to decide whether or not it is in the best interests. Whereas we said, "Well, that's all very nice, but we don't have enough faith in commercial interests to feel comfortable with that," so we look at having an additional step that considers market impact in the registration process because all the other factors in the registration process look at non-market factors. (SK#18A, SARM)

As noted by SARM, the biotechnology industry is directly opposed to any form of regulatory solution, as they feel the regulatory system is already overly burdensome. Ag-West Bio, for example, argued that the regulatory system "sometimes deals with emotion and politics rather than science, and there is a vocal minority that wants to move it in that direction." They wanted to make sure that it "remains science-based and just looks after science-based regulatory analyses" (SK#27A, Ag-West Bio, Inc.). In response to the problems with RR wheat, Ag-West Bio remained adamant that market acceptance should be kept out of the regulatory system, and decisions on introductions should be made only on the basis of scientific parameters: "[RR wheat] is a market acceptance issue, and so the industry has to deal with that. And I think in Canada the industry has demonstrated that they can deal with it effectively" (ibid.).

Those concerned about their livelihoods and dependent on garnering sufficient lobby pressure to preserve it clearly feel that the monumental effort required to stop RR wheat did not indicate an "effective" means of dealing with the issue, however, and they continue to strive for more input in regulatory approval. Despite the obvious power and control issues manifested in the RR wheat conflict, the issue was not the harbinger of an anti-biotech sentiment among producer organizations that it might have been. Objections remained firmly market based:

If consumers supported it, we would support it. (SK#24, CWB)

If anything, the majority of these organizations were careful to couch their opposition in the specifics of the case, and not to taint the technology itself. The sense that no one wanted to compromise the promise of biotechnology for the future of agriculture in Saskatchewan was palpable. The dynamics of resistance here are thus very different from those that arose in response to the Schmeiser case.

Give Us Back Our Genes: Patents on Life, Monsanto's Genes, and Schmeiser's Canola

The Schmeiser case started in 1998. The first trial revolved around whether Percy Schmeiser, a canola grower in Saskatchewan, had infringed on Monsanto's patent by using its RR technology without contract. Mr. Schmeiser alleged to the contrary that Monsanto's unwanted genetic material had contaminated his crop, and that he faced a serious loss of rights on account of Monsanto's claim. The case went all the way to the Supreme Court, where the broader issue of the patentability of life came under question. In addition to defending against Monsanto's patent infringement lawsuit, Schmeiser launched an international public relations campaign against the company and against biotechnology more broadly that was still ongoing ten years later. He alleged that self-propagating patented technology violates farmers' rights, as the technology establishes itself on private land and then is subjected to industry's ownership claims; that Monsanto's investigative team — Robinson Investigation — is intimidating farmers into accepting unfair settlements to avoid costly infringement litigation; and that the issue of gene flow has turned farmer against farmer and resulted in a culture of fear on the prairies. In rocketing these issues into the public forum, Schmeiser highlighted the potentially devastating power shift that patents on self-reproducing seeds could produce. The full legal details of the case will be discussed in Chapter 5; however, enough details are provided here to consider the impact of the issues on the community.

While Schmeiser suggests that a whole package of changes has been visited on the prairies as a result of GM technologies, he clearly has a vested interest. To what extent are other farmers impacted, or even aware of these changes? Given that when RR canola was first introduced, Monsanto required producers to attend a meeting informing them of all facets of the new technology, most producers should have been well aware of restrictions on seed saving and the other key aspects of Monsanto's contracts. For those who weren't, the ensuing years of media coverage around the Schmeiser and Hoffman cases fully vetted the issues. Consequently, there was an unsurprising awareness of the proprietary aspects of agricultural biotechnologies and, more specifically, of patent infringement, in Saskatchewan. Further, there appeared to be few who were not thoroughly aware of the Schmeiser case, and fewer still who did not have an opinion on it, sometimes a strong opinion. Even retrieving court documents in Saskatoon got me an earful from a clerk about "that man." Outside of those who endorse organic or sustainable agricultures, these opinions frequently fell against Schmeiser, although the more thoughtful would often note the difficult ethical issues involved. For those who had negative opinions, Schmeiser was sometimes reported as someone who likes to be in the thick of things, to the point of creating the thick to be in the thick of:

> His neighbors and some of the people in his area are not surprised that he got in trouble. And, you know, they say he's always done those sorts of things, and always poked the hot stick in somebody's eye, and unfortunately he poked it in Monsanto's eye, and they didn't take it lightly! (SK#14, GM producer)

However, while some had plausible explanations regarding how they arrived at this information, Saskatchewan isn't small enough for so many to have formed direct personal opinions of the man. Indeed, many of the personal opinions sounded overly similar. Likewise, details of the case were often cited uniformly incorrectly when used to criticize Schmeiser. For example, the lawsuit involves supported allegations that—whether he originally obtained them by natural processes or not—Schmeiser saved seeds that contained the GM technology, and then reseeded them the following year, producing a regular crop. Nonetheless, I was repeatedly told that Schmeiser's story of contamination was implausible because there was no way the GM seeds could have blown into his fields in such "nice straight rows," neglecting this point of reseeding. Similarly, while it is the norm for legal arguments to shift as they move through levels of court, and despite the fact that Schmeiser did not change his claims on the basic facts of the case (although Monsanto did drop their allegations of brown bagging), it was with some indignation that Schmeiser was reported to have continually changed his story from denying factual guilt (as would be appropriate at the trial level) to trying to challenge the validity of the patent itself (as would be appropriate for the Supreme Court, which does not retry facts). The consistency of responses suggests that the counter-message to Schmeiser's campaign was having important effects.

Given these qualifications, opinions about the legitimacy of the case among GM producers and related stakeholders were fairly consistent. The following are a few typical samples:

> As far as I'm concerned the whole thing is a farce. . . . there's no way that that amount of seed would blow and contaminate onto his ground and contaminate all his canola acres. It's just not going to happen. (SK#29, GM producer)

> People around here are a little bit skeptical: they wonder how it came to pass that the Roundup Ready canola came to grow in nice straight rows in hundreds of acres, like it didn't seem to be a random thing; it seemed to be more of a deliberate use of Monsanto's seed. (SK#3, seed dealer/producer, GMO)

> If I had a whole field coming up RR canola, somebody seeded it, it didn't all blow there. (SK#25, GM producer)

There are, of course, significant reasons why Schmeiser would not be considered a hero among canola growers in the region. Notably, his rise as an anti-GM spokesperson over the long duration of the case (and continued in a small claims court action) tempered the support many felt, as will be discussed. While a number of people expressed either skepticism or exasperation with Schmeiser, however, this perspective was by no means unanimous. Those who were more inclined to be supportive were less concerned with the practicalities of infringement and more concerned with the issues behind the case. Further, and not surprisingly, a large contingency of this support could be found in the organic sector, as exemplified by the following:

> The ruling that Percy Schmeiser was responsible for what had blown onto his land—it's just totally bizarre in my opinion. (SK#15, organic producer/retailer)

Schmeiser was growing whatever he was growing in a regular style commodity and then . . . was contaminated with a genetic canola, so it wasn't
his fault. Like he has no control on it, and until it's proven and established and all the flags, if you like, are taken out of the problems, it
shouldn't be that it's forced on somebody else. . . . it shouldn't be a problem for me. Just because my neighbors are doing something, it shouldn't
be my fault or it shouldn't affect me. It shouldn't have a bearing on what I
do. (SK#13, organic producer)

The more sympathetic treatment of Schmeiser by organic farmers is
telling in that organic producers have little to lose from the negative press
that Schmeiser has generated and significantly more to win by the iron-
ing out of the liability issue. Nonetheless, it was not strictly organic producers who felt some support for the issues behind Schmeiser's claim. For
example, one conventional producer, who mainly grew wheat, hay, and
crops other than canola, expressed his feelings about the matter this way:

But if it was up to me, I would have still found him not guilty. Even if he
was guilty, I would have still found him not guilty! . . . if you are going
to mess around with this stuff, and spread it around in people's fields, it's
your problem. It's the company's problem. If they insist on a patent right
to the use of all of this stuff, then they are also required to ensure that
none of it contaminates into the neighbor's fields. . . . I have no doubt
that there was some of this GM stuff that accidentally fell into Schmeiser's field. I'm quite sure that that probably did happen. Maybe not to the
extent that he was claiming, but nonetheless, the fact that it was there in
however minor quantities absolved him. (SK# 33, conventional producer)

Thus for some, the issue of liability around self-reproducing "inventions" is more important than the practicalities of the case itself, although
they clearly remain in the minority. It is also important to note that even
some of those who were less supportive of Schmeiser readily acknowledged the possibility of genetic contamination:

To me, yeah, he could have gotten some contamination from a truck
going down the road or whatever. There's combines going down the road
at harvest time, who knows, grain elevators, everything's kind of drip-
ping and dribbling all over the place. There's a little seed all over the
country. (SK#14, GM producer)

The objection was to the extent of contamination, which was alleged to be over 95% GM by Monsanto, and which the trial judge found to stand up as fact. Given this percentage (albeit disputed by Schmeiser), many felt that there was no doubt about deliberate infringement, and few expressed concern over the possibility of farmers being unjustly sued for infringement. Even the GM producer just quoted, for example, felt there would be a legal difference between a full GM crop and a crop that was partially contaminated:

> I'm sure if they found a little Roundup Ready on the edge of my field, but found that the rest of the crop was a conventional. . . . then I'm sure if they wanted to take me to court then the judge would laugh them out of the door. (SK#14, GM producer)

Even for those who were aware of the legal gray area that remained after the Supreme Court ruling left undetermined the percentage of GM material that constituted infringement, the general sentiment appeared to be that Monsanto would be foolish to pursue anyone who was legitimately contaminated, and that the company would only take legal action against those who deliberately infringed. While the occasional farmer entertained the possibility of a bias against Schmeiser, most expressed that the legal framework was likely not a problem for the average farmer:

> It is too bad, if you did get accused and you weren't actually using it, because the seed is blowing around all over the place. It may be hard to prove in some cases, although if you have a crop that is 99% Roundup Ready, it is hard to argue that I guess. (SK#3, seed dealer/producer, GMO)

> I think [companies] treat farmers fairly. Farmers abide by—when they buy the seed and they pay for it up front—when they abide by the rules, then there's just no way to get into a jam. (SK#29, GM producer)

An important aspect of the legal framework is that producers who are accused of infringement are faced with either proceeding through expensive litigation in their defense or signing a settlement (including a significant fee) with Monsanto. A key feature of the settlement option is that it is accompanied by a nondisclosure agreement; thus if producers were having a problem, they could not communicate it to their neighbors or broader community. Producer organizations which would likely field complaints, if only anonymously, were similarly unconcerned about the interaction between farmers and Monsanto over these agreements, however. According to the Saskatchewan Canola Development Commission, nondisclosure agreement or not, you would hear about it if farmers were being bullied:

> If it was a legitimate problem there would be lots of farmers raising hell. Farmers are not the kind to not do that. (SK#28, SCDC)

The SCDC was not the only organization to think the issue of farmer liability wasn't significant. Some organizations, like the Saskatchewan Canola Growers Association [SCGA], are directly responsive to their members, such as through conventions. While acknowledging potential issues, they still found no indication of a practical need for involvement in the issues around Schmeiser's case:

> But if the shift changes where more people are starting to disagree or they feel one of these companies are getting heavy handed with them, then it could swing in the other direction, and as an organization we'd have to look at it more. But until that point, where our membership is generally happy with the situation, then it's not a problem for us. (SK#21, SCGA)

Similarly, the Saskatchewan Association of Rural Municipalities stated:

> I think that the Schmeiser case was a unique case and we've only seen that one case with that particular problem. If we had one case one year, and the next year we had two cases, and then we had five cases, then I think it would be something we'd put up our antennae and say, "OK, why is this happening and what's going on out there?" (SK#18A, SARM)

In reality, the issue of GM contamination in canola practically resolved itself with the rapid speed of GM adoption. With the very small percentage of conventional canola remaining—and the vast majority of that regionally specific—there are few producers left who could be contaminated. Consistently, the interviews here provided no suggestion that the threat of liability pushed any producers to adopt biotechnologies as an evasive maneuver, or that farmers in general felt at risk of liability for contamination. Nonetheless, the importance of the issues raised in the Schmeiser case remain, as with each new GM introduction they will reignite—particularly in any crops that do not enjoy the same speed of adoption. Thus, the actual long-term impact can be a very different thing, and the issues were aired sufficiently that one might expect a more empathetic treatment of anyone willing to pursue them through years of litigation. I will return to this point after addressing Saskatchewan's second distinguished biotechnology-related lawsuit.

Take Back Your GMOs: Genetic Contamination in the Organic Industry

Organic producers provided the other main source of resistance to biotechnologies in Saskatchewan. Organic production is a growing agricultural niche in Canada, particularly in Saskatchewan. As noted, by 2003 organic producers made up 2% of the province's producers, and their numbers were growing. Certification for organic production includes such requirements as a prohibition on the use of synthetic chemical inputs, buffer zones against noncertified crops, and for the land to have been free of these inputs for several years before certification is granted. Genetically modified organisms are considered a synthetic input and are banned in organic production. In return for the stringent requirements of organic production, and the additional record keeping and auditing requirements, organic producers receive a significant premium for their product. Given the usually smaller acreage and the greater effort involved—including the long period of chemical-free farming prior to certification—a loss of certification would be a devastating economic blow to an organic producer.

Organic crops in Saskatchewan are mainly grain crops produced for export, the principal markets being the United States, Japan, and Europe; thus these crops must also meet the organic standards of those markets. Standards relating to the use of GMOs were slowly put in place in various countries following biotechnologies' commercialization. While GMO contamination thresholds and their dates of implementation have been disputed in the courts, the fact that GMO contamination is currently a threat to organic certification is not in question. The Saskatchewan Organic Directorate [SOD] claims that since the introduction and widespread proliferation of GM canola, it has become impossible to obtain GMO-free seeds and to produce GMO-free canola. Hence, they claim that the production of organic canola is no longer possible in Saskatchewan, and organic producers have consequently lost both an important market and an important crop for their rotations. Were canola to be the only crop removed from organic production, the industry might not have been spurred into action. With the attempted introduction of RR wheat, however, organic farmers were beginning to see the writing on the wall for their industry.

While not as well publicized as the Schmeiser case, the SOD case (technically, the Hoffman case) is the first example of a group of producers going on the legal offensive against a biotechnology company. It was initiated when two organic farmers applied for class action status over the contamination of their organic canola and simultaneously filed an injunction against the release of RR wheat. The injunction application was dropped when Monsanto withdrew its application for RR wheat, and the case proceeded on the issue of GM canola alone. The details of the case will be discussed in Chapter 5, but at its heart is the concern that organic producers be able to continue to produce organically and to receive a premium for their product.

Unlike in the Schmeiser case, the concerns of the representative farmers in the SOD suit can be easily found replicated throughout the organic community. At its simplest, organic farmers are concerned about the loss of canola as an organic crop. Though the organic canola industry was still in its infancy when GM canola was introduced, relatively clear examples of the impact of GM canola are nonetheless available. As one producer stated:

> You know, there's no organic canola around here. It's basically impossible to grow it because there's too much contaminated land. (SK#8A, organic producer)

While outright contamination was a clear and present danger, producers also suffered from market rejection over the assumption of contamination. One organic producer who also operated a sprout business selling to the health food industry explained how within the first few years of GM canola's introduction they lost about 75% of their canola market:

> Canola is really actually a nice sprout, but it's just like one of those things that got labeled. I mean, there is non-GM canola out there, right, but canola has got such a bad reputation in [the] consumer's mind. (SK#15, organic producer/retailer)

Markets such as in Quebec and Europe were unwilling to buy their canola products, which forced them to cease production of what used to be an important crop on their farm.

From outside of the organic community, the response to the concerns of organic producers was at times empathetic, at times self-righteous. Producers—many of whom weren't familiar with the specific details of the case—were more inclined to the former (empathizing with any impact on a grower's income). Producer organizations and other stakeholders were far more inclined to the latter. From these there was a fair amount of clucking over the organic industry for setting a standard that could not be met, and consequently putting themselves in direct opposition to the rest of the (GM-based) canola industry:

> I think they've drawn the line in the sand far too pure. There was room for organic and GMO to coexist and go their merry way. (SK#22, agricultural consultant/media)

> Personally I don't understand why [organics have] gotten themselves in that box. Because they are essentially guaranteeing that there is no GMO present in their product. Zero. They don't make that guarantee for pesticide residues. (SK#2, academic, crop development)

For those who held this perspective, resolution lay in simply creating a GMO tolerance limit that organic producers could meet, such as allowing certification with no more than 1% GM content, for example. By not doing so, organic producers were seen to be unnecessarily pitting themselves against GM canola producers, although some are more willing to acknowledge that it might not be up to organic producers themselves so much as their markets. A number of non-organic producers and stakeholders also argued that there really was no organic canola industry to speak of in Saskatchewan, and that whatever industry there was couldn't be sustained anyway, given the weed challenges of canola production. Some argued further that weeds spreading from organic farms represented its own form of contamination, and that the argument could go both ways over who was contaminating whom. At the heart of most of this anti-organic sentiment seemed to be a simple principle of self-defense. The organic producer's lawsuit directly challenged the makers of a technology that the majority of canola growers valued and wished to see maintained:

> What do you want us to do? . . . Can you tell everyone that produces 10 to 12 million acres [to stop] because there's a few hundred acres out there that want to be organically grown? . . . These people that talk this way, I'm afraid it's a small group, which certainly has a right to grow, by

all means, and has an issue of contamination. I'm not saying that can't happen, and won't happen, and isn't happening; I'm sure it is. However, to restrict how the whole industry, that you have to bend because of my little group, you're not going to have that. That's not going to happen. I'm sorry. (SK#4, Canola Council of Canada)

Similarly, a representative from another agricultural organization who wanted his personal opinion disassociated from his organization, characterized the situation thus:

I think that's sort of a failure on their part, to expect a zero tolerance, and expect us to sacrifice basically a billion-dollar industry on them. I think at their peak, when they were growing canola, was less than a million dollars net return to them. I think sometimes the majority has to rule the situation, when we are adding that much value to our industry, to build a billion-dollar industry. (confidential)

From an organic farmer's perspective, however, the problem of market confidence may not be so simple to solve with the introduction of achievable tolerance levels, even assuming such levels could be agreed upon in various markets. Further, the concern of organic producers was far greater than the issue of GM canola. Ironically, their greatest call to action was over a concern they held in common with many GM canola growers—the intended introduction of RR wheat:

If they brought GMO wheat in, we could be in really serious trouble because wheat is more along the lines of grass and can cross-pollinate and some other things like that and then we could lose . . . everything. (SK#8A and SK#8B, organic producers)

Like if wheat, if they came up with a Roundup-Ready one—whatever— and it was getting into the organic stuff and they couldn't keep it out, I think that would really hurt a lot of the organic people. (SK#23, organic producer)

Whereas GM canola was introduced while the organic canola industry was still developing—thus representing a small actual but large potential impact—GM wheat would affect a large number of farmers already producing organic wheat. Significantly, a number of these farmers felt that the introduction of RR wheat could spell the end for organic farming.

It would, as one organic producer bluntly stated it, "knock us out of the organic business" (SK#12). According to another:

> GMO wheat would be really hard because it would take away one of the most—the main organic crops. Once everything is contaminated with that . . . Losing organic canola was a big loss. Losing organic wheat—I'm not sure the farmers could hang on. (SK#8A, organic producer)

While the withdrawal of Monsanto's application for RR wheat brought many a sense of relief, it was by no means an end to their concerns. Genetically modified alfalfa was the next crop on Monsanto's slate for release. Alfalfa is highly important to crop rotations, especially for organic production, because it is a nitrogen-fixing crop. Its removal from organic farming would mean the loss of an important soil-improving tool and of alfalfa hay as a feed for certified organic livestock production. Once again, organic producers faced the potential loss of a significant tool for their livelihood.

In response to the question of what he would do if GM alfalfa were released, one producer stated:

> Pray. I don't know to be honest with you, because we need alfalfa to feed our animals, you know, on our crop. So if there's GMO alfalfa across the road, that's when a guy is going to have to step in and say to the neighbor, "Joe, like, don't be planting that out there." I don't know. I'm not sure. (SK#7A, organic producer)

Similarly, the organic sprout dealer who lost her canola market predicted that GM alfalfa could represent a 30% loss of the volume of their sprouts. As she explained, it wasn't that they couldn't replace alfalfa with another crop, but the trend would lead to the demise of their business:

> I mean other legumes can be used. But part of it is, if alfalfa is introduced and then it's another crop and another crop. (SK#15, organic producer/retailer)

In short, the creeping introduction of biotechnologies caused significant concern for organic producers. Thus, the Hoffman case did represent concerns held by many organic farmers. As seen with RR wheat, an enormous and broad-based lobby effort was necessary to stop the introduction. Not only is achieving such broad-based opposition improbable for organic farmers, but even within the sector, maintaining constant opposition is difficult. Despite the vested interest, for example, there was difficulty raising a sufficient resistance to lobby against GM alfalfa even within the organic community. Some saw this as part of Monsanto's strategy:

> And that's what I'm kind of thinking—is they're just keeping on developing other crops hoping that some way they will just sneak through without people—like get more and more crops—that's their game plan or whatever. Weasel low for now and go in there and get that one in, and that one in, and that one in. (SK#15, organic producer/retailer)

A lawsuit where producers can have their concerns represented would circumvent this need for constant lobbying, however, and consequently many organic producers were glad to contribute to the fund set up by the SOD for the lawsuit. Unfortunately for these producers, their class action suit could not be amended to include an injunction against RR alfalfa, and opposition was not sufficient to prevent its approval. In 2005, RR alfalfa was approved by both the CFIA and Health Canada for release, although by 2009 it was still pending registration for commercialization (see Chapter 5).

The Hoffman case was thus dual purpose. On the one hand, it provided a means for organic farmers to fight for their cause without needing to maintain a constant and exhausting full-scale lobby effort. On the other hand, it provided a means to legally challenge the "ownership without responsibility" privilege that biotechnology companies appear to have captured with their patents on life. As one organic producer put it:

> I'm hoping that their responsibility for it—for their rights—will be reined in to the point where they will be responsible as well. They have all these rights but they're not living up to their responsibility on the other side of the coin, is my feel on it, and I think if they were held responsible for the damage they do, they might disappear real quick. (SK#8A, organic producer)

For some organic producers this is strictly a market concern. For others, the concerns around biotechnologies incorporate a wide range of potentially devastating environmental repercussions. In either case, the drive to force legal liability for GM contamination aims to cut to the industry's heart by challenging its externalities, and attempting to make them part of the biotechnology equation. At the very least, such legal actions have the potential to create instability in biotechnology investments.

Opposition and Its Opposition

As RR wheat and the Schmeiser and Hoffman cases have shown, there have clearly been moments of opposition to biotechnologies in Saskatchewan, each triggering a fair amount of publicity and support, even outside the province. What the impact of this opposition has been on the industry is hard to quantify or even fully qualify; it is certain, however, that there has been an impact. The RR wheat conflict, for example, may have reduced research and development into new biotechnology applications—speculations abound that as a result of the altercation Canada lost some of its favorable status for technological development:

> Canada has actually been a fairly—I guess you can say—hotbed of bio-technology in crop agriculture and development in new crops, and I don't know officially the numbers, but I have heard that research dollars that would be sent to Canada have decreased because of the Roundup Ready wheat issue: because they no longer look at Canada as a favorable place to develop these products because of the resistance fueled, and that may have hurt the amount of research being done here. (SK#21, SCGA)

> You can already see the implications of the uncertainties of commercial returns to the companies, and that is that they don't do as much as they used to. Five years ago I would talk to the companies and they would say they are working on ten different technologies in fifteen crops; now they are working on three technologies in two crops. (SK#16, academic, agricultural economist)

Whether Canada has actually been penalized in this way is hard to ascertain, although it is clear that the industry cannot take the commercialization of new biotechnologies in Canada for granted. While RR wheat exploded into a public relations nightmare for Monsanto, it was not necessarily an isolated incident, as the earlier withdrawal of GM flax attests.

With respect to the two lawsuits, the impact again is impossible to ascertain, although it is evident that the publicity around the cases, and the associated negative environmental coverage of biotechnology, has contributed to greater consumer sensitivity and consequent pressure on the industry. As one knowledgeable informant characterized it, when GM technologies were first introduced, and industry was attempting to garner excitement over its product, consumers "could give a rat's ass." All this has now changed. The negative attitude of Europeans, the efforts of environmentalists, market risks, and other factors have all contributed to this change, and there is little doubt that the publicity generated around Saskatchewan's two court cases has also contributed.

The Hoffman case went far to publicize the impact of biotechnologies on organic producers, and move questions of liability to public debate, if not yet policy debate. The impact of the Schmeiser case can be to some extent extrapolated from the negative treatment he sometimes receives in his own community. Although there is no doubt Schmeiser has support, it is also plain that the Saskatchewan agricultural community has not circled the wagons around him, even though he would appear to be championing the case of the little guy against the big companies. To some extent this is the trickle down of negative press from invested organizations. For some, Schmeiser lost credibility when he pursued Monsanto in small claims court, alleging the company was liable for contaminating his wife's organic garden. A likely even more significant factor, however, is that in his ongoing crusade against Monsanto, Schmeiser allied with environmental organizations and spoke out internationally not just about GM's legal issues, but also criticizing the health and safety of biotechnologies—including GM canola. The Saskatchewan canola industry was trying to position canola as a healthy alternative to cheaper oils on the market, so Schmeiser's well-publicized attacks on biotechnologies were a direct affront to those concerned about canola's marketability.

There is ample evidence of resentment over this issue. For instance, concerns over the negative press around GM crops prompted members of the Humbolt and District Marketing Club to wryly suggest suing Schmeiser for the losses they had incurred ("Consumers," 2001). Similarly, a respondent from the Saskatchewan Canola Growers Association commented that while the case didn't warrant the attention it gained, it did manage to put a bad light on canola:

> I don't think it has, like, reduced the price of my canola, but it has hurt the industry as a whole to some extent. (SK#21, SCGA)

This impact is arguably not strictly limited to canola; the negative press may have also affected the introduction of new GM crops:

If they could get market acceptance they could have Liberty Link peas tomorrow. (SK#28, SCDC)

Perhaps best summing up the development of biotechnologies in this controversial context is the response of an academic in crop development when asked his opinion on the social impact of biotechnologies in Saskatchewan. He replied that while it had the potential of being positive, the current reality was just increased stress levels for many people. Given the volatility of public opinion, the end result of this battle is far from determined. A significant advantage on the side of industry is the fact that growers are very conscious of their economic vulnerability and keen to avoid disadvantaging themselves along these lines. While more explicit on the point than many, the sentiments of the following producer resonated with concerns expressed by others:

If we choose not to grow a certain product because it's a health risk, maybe another country will, and [will] say, you know, "We're not concerned about it," and the same multinational company then can farm its resources to a different nation. You know, that's entirely possible. So, you know, I'm thinking we're going to work in this area. (SK#11, GM producer)

Thus, just as surely as there is opposition to the technology, which has an effect on its development, there is also a significant force of opposition to the opposition.

Conclusion

Based on the preceding two chapters, conclusions about whether there is an expropriationist tendency as a result of the introduction of GM technologies are best drawn in terms of short- and long-term perspectives. In the short term, biotechnologies have definitely provided a means to assist farmers to stay in the agricultural game. In the long term, there are equally definite indications of a trend of expropriationism. In addition, some unexpected control issues arose with respect to corporate introductions of unwanted GM crops. As biotechnologies are introduced into more crops, as commercially viable non-GM alternatives of these crops decline, and as prices and contract restrictions increase, farmers will have less and less recourse, and expropriationism will inevitably increase.

Many of these tendencies are already at play in agriculture more broadly, through the growing prevalence of production for contract and the declining public sector involvement in agriculture, but biotechnologies take them further, faster.

Of course, opposition to the industry, and to its expropriationist tendencies, may have an impact on the future of biotechnologies. The impact of this opposition is still unfolding. Contamination of organic production is likely to be a continued source of opposition, for example. Efforts to introduce market considerations into the technology approval requirements were another significant modifying factor. Beyond the marketability issue, few outside of the organic industry are in favor of tightening the regulations around biotechnology; a number even felt that such a move would only decrease the participation of the public sector in plant breeding, one of the few areas where growers felt the government should have greater involvement. In the following chapter, the Hoffman and Schmeiser lawsuits will be considered in more depth.

CHAPTER 5

Legal Offense and Defense
on the Canadian Prairies

They can patent anything they want, design and engineer anything they want, without responsibility. And even have a Technology Use Agreement, where they maintain control, and maintain the financial benefits of the ownership, [and] they don't have to take any of the responsibility for how the person they've licensed to use it handles it.
SK#6, PRESIDENT, SASKATCHEWAN ORGANIC DIRECTORATE

You know, it's their property when it comes to a farmer potentially having to pay, but it's the farmers' property when it comes to them having to potentially pay, and we don't think that they can take those inconsistent positions.
SK#1, LAWYER FOR SASKATCHEWAN ORGANIC DIRECTORATE AND SCHMEISER

You don't know how strong that patent really is until somebody violates it and it's upheld in a court of law.
TRISH JORDAN, SPOKESPERSON FOR MONSANTO, CITED IN LYONS, 2001

Introduction

As we saw in Saskatchewan, biotechnologies have technological novelty; they also introduce a host of new ownership and control issues to be contended with at the farm level. Both Percy Schmeiser and the organic farmers have gone far to publicize the legal issues that accompany their introduction. On a practical level, these issues remain theoretical for many producers, whose technology decisions largely result from economic cost-benefit assessments. As demonstrated by these lawsuits, however, the legal changes associated with biotechnologies represent a bit of an iceberg phenomenon: while the shift to proprietary seeds is apparent and navigable (by not saving seeds), a number of changes remain below the surface. This does not reduce their impact, but does delay producers' awareness of them. In order to assess the nature and direction of any reorganization of agricultural production facilitated by biotechnologies, therefore, it is necessary to consider the legal changes evolving below the surface—most notably, provided here in the legal details of Saskatchewan's two groundbreaking court cases.

This chapter will first look at Canada's intellectual property rights [IPR] protection for plants, both through legislation enacted under international obligations and through the evolution of Canadian case law. Next, it will provide a more detailed assessment of the Schmeiser and Hoffman lawsuits—with a number of questions in mind. What are the resolved and the unresolved legal issues around agricultural biotechnology in Canada? To what extent are any legal resolutions facilitating a reorganization of agricultural production? Who gains and who loses in this reorganization? In what direction do unresolved issues seem to be unfolding? This investigation aims to uncover not only whether there is evidence of an emerging trend of capital accumulation through legal means (or "expropriationism"), but also the extent to which there is resistance to it.

The data for this chapter are made up of the court decisions and supplementary legal documents of the selected cases—such as court transcripts and affidavits—as well as interviews with litigants and their representatives, wherever possible. More detailed information on interviewees is available in the appendix. Monsanto Canada preferred to speak through a single representative, their communications officer, Trish Jordan. For this reason, the perspective of Monsanto's legal representation was not available. Despite agreeing to respond to questions on condition they were submitted in print, Bayer CropScience was unable to respond within the available time.[1] Further, although a lawyer for the company indicated a willingness to be interviewed, this was not permitted by the company. The Schmeiser case was concluded at the Supreme Court of Canada in 2004. While the Hoffman case, initiated by the Saskatchewan Organic Directorate, could still have proceeded through individual claims once class action certification was denied in 2007, the directorate's April 2008 decision not to proceed effectively brought its conclusion.

Intellectual Property Rights Protection

As noted, the WTO's TRIPS Agreement seeks to impose uniformity in intellectual property protection—if not providing full patent protection, member countries must have a sui generis system of intellectual property protection, such as that modeled by the International Union for the Protection of New Varieties of Plants [UPOV]. Canada enacted the Plant Breeders' Rights Act [PBR Act] in 1990 and became a signatory to UPOV in 1991, although it is party to the 1978 act of the convention. Plant protection under UPOV 1978 provides breeders with rights over production for purposes of commercial marketing, the offering for sale, and the ultimate marketing of the variety (UPOV, 1978: Art. 5). At the same time, it provides exemptions for the rights of researchers and for farmers to save seeds for their own use.

While the act explicitly mentions only the exemption for researchers—who are permitted to "[utilize] the variety as an initial source of variation for the purpose of creating other varieties or for the marketing of such varieties" (ibid.)—the exemption for farmers (the "farmer's privilege") has been widely adopted. According to the International Union for the Conservation of Nature and Natural Resources [IUCN], "the limitation of Plant Breeders Rights to production for the purposes of commercial marketing etc, has been interpreted in practice as allowing farmers to re-plant and exchange farm-saved seed" (IUCN, "Article 9"). The Canadian Food Inspection Agency [CFIA] states that in Canada, "[f]armer's privilege was allowed in the current PBR Act because support for the legislation from some farm organizations was conditional on allowing farmers to retain the right to save and use their own seed" (CFIA, "Proposed Amendments"). Whatever its source, the provision for farmer's privilege under the 1978 act is undisputed in Canada.

In 2003 a number of industry groups[2] under the auspices of the Canadian government initiated an industry-wide assessment of the Canadian seed sector, called the Seed Sector Review [SSR]. The purpose of the review was to assess the regulatory environment and develop recommendations for regulatory change. One of the recommendations of the SSR was to upgrade to UPOV 1991. In contrast to the 1978 version, UPOV 1991 explicitly makes these exemptions optional features to be adopted or excluded at the discretion of the signing country:

> [E]ach Contracting Party may, within reasonable limits and subject to the safeguarding of the legitimate interests of the breeder, restrict the

breeder's right in relation to any variety in order to permit farmers to use for propagating purposes, on their own holdings, the product of the harvest which they have obtained for planting. (UPOV, 1991: Art. 15[2])

While those in favor of UPOV 1991 argue that making this provision optional does not mean that it will be excluded, those opposed argue that opening the door to its being optional renders its actual exclusion simply a matter of time. There is little doubt for those in favor of farm-saved seed that UPOV 1991 is a threat. The upgrade to UPOV 1991 is only one of the SSR's suggestions, however. The overall thrust of the review is "streamlining" regulations and facilitating opportunities for industry, many aspects of which again raise the concerns of those in favor of farm-saved seed.

Another erosion of farmers' right to save seed previously afforded by UPOV 1978 and the PBR Act occurs through the introduction of seeds available only by contract. In fact, this erosion appears to be facilitated by Agriculture and Agri-food Canada [AAFC]. Agriculture Canada applies for PBR protection on the seeds they develop and then licenses them to seed companies for multiplication and distribution. The licensees can then introduce contracts into the equation. Hard White Wheat, for example, a variety developed by AAFC and licensed to Quality Assured Seeds, can only be purchased under contract conditions that preclude seed saving, and that make a "mockery" of the PBR provision for seed saving (Beingessner, 2004: n.p.). The argument for such a system is that it provides funds for research: against it is the argument that "forcing farmers to buy new seed each year benefits only those . . . that sell the seed" (ibid.).

This issue of seed-saving rights is not just playing out in government legislation, but is intimately connected to court decisions regarding the patentability of life. As we will see later, this question had an early start in the United States, with a trend supporting the patentability of life strongly set by the 1980 Chakrabarty case. In Canada, the transition to proprietary life forms has proceeded more slowly. According to Roberts (1999), given the lower rates of Canadian versus U.S. investment into commercial biotechnology, both the Canadian Intellectual Property Office [CIPO] and the Canadian courts "have faced much less pressure than their American counterparts to accede to an expansive view of patentable subject matter" (30).

In order to obtain patent protection in Canada, an invention must have novelty, utility, and ingenuity (CIPO, "A Guide"). A patent grants its holder the right to exclude others from making, using, or selling the invention for 20 years. In return, the inventor is "expected to provide a full description of the invention so that all Canadians can benefit from this advance in technology and knowledge" (ibid.). Simply stated, patents are intended for the promotion of innovation for the benefit of society. According to the Patent Act, the definition of invention is "any new and useful art, process, machine, manufacture or composition of matter, or any new and useful improvement" in such (Government of Canada, *Patent Act*, R.S., 1985 c. P-4, s.2). This definition is modeled on U.S. law (Vaver, 2004:157). In contrast to the apparently enthusiastic expansion of the definition of "manufacture" and "composition of matter" evident in the American approach, particularly in the last two decades, the legislative history of Canada has evolved a somewhat more restrictive definition (Atkinson, 2005:13).

A Canadian Patent Appeal Board decision regarding a mixed yeast culture designed to purify pulp mill effluents in the 1982 *Re Application of Abitibi Co.* (hereafter *Abitibi*)[3] represents the board's first decision in favor of patenting life forms (Roberts, 1999:31).[4] While the patent examiner originally rejected the claim on the basis that life forms are not patentable, the appeals board noted that judicial bodies throughout the world have "gradually altered their interpretation of statutory subject-matter to adapt it to new developments in technologies," and concluded in favor of patenting life forms modified by human ingenuity (*Abitibi* at 88–89, as cited in Roberts, 1999:31). The board stated that it could "no longer be satisfied that at law a patent for a microorganism or other life forms would not be held allowable by our own courts" (ibid.). It further indicated criteria for the future of such patenting, finding that lower life forms—such as yeast, moulds, fungi, bacteria, and the like—that can be mass-produced in a uniform manner should be patentable. While acknowledging that such uniformity would be difficult to achieve in higher life forms, the board did not rule out their patentability should they eventually meet these criteria: "[I]f it eventually becomes possible to achieve such a result, and the other requirements of patentability are met, we do not see why it should be treated differently" (ibid.).

Roberts notes that if the courts endorsed the approach outlined in *Abitibi*, it might have had significance akin to that of the precedent-setting Chakrabarty case in the United States (see Chapter 7). As it is, patenting life forms was substantially qualified in *Pioneer Hi-Bred v. Canada* (*Commissioner of Patents*) (hereafter *Pioneer*).[5] *Pioneer* involved a variety of soybean produced by selective breeding, the patent for which was denied by the Commissioner of Patents on the basis that it was not an invention under the meaning of the Patent Act. The federal court of appeals concurred, indicating that while special legislation should be adopted to provide some intellectual property protection for plant breeders, if Parliament had intended patents on plant varieties, it would have provided for it through the inclusion of applicable terminology—such as "variety" or "strain"—in its definition of invention (Roberts, 1999:33). *Pioneer* was appealed to the Supreme Court of Canada, where the patent application was rejected, but on the basis of inadequate disclosure[6] rather than over the patentability of life. In its obiter dictum, however, the court distinguished between GM plants and those resulting from selective breeding, as only the latter rely on evolution and the laws of nature, stating that "it was more likely that genetically engineered plants would be patentable because they resulted more from human intervention and less from the laws of nature" (Atkinson, 2005:14).

The PBR Act was passed soon after, in 1990. Based on these court cases, the Patent Office issued guidelines for patent examiners, differentiating lower life forms ("which are essentially unicellular in composition") and higher life forms ("which are multi-cellular differentiated organisms [plants, seeds and animals]"), stating that "[l]ower life forms which are new, useful and inventive are patentable," whereas "[h]igher life forms are not" (*Canadian Manual of Patent Office Practice*, cited in Atkinson, 2005:15).

Whether life is patentable in Canada would seem to have found final legal expression in the much-celebrated case of the "Harvard mouse," or "oncomouse," a mouse genetically modified to be predisposed to cancer, useful for research purposes. The mouse had already been patented in the United States in 1988, but based on its status as a higher life form, the Commissioner of Patents in Canada considered it unpatentable subject matter. The denial of the patent worked its way through the Canadian courts all the way to the Supreme Court of Canada in *Harvard College v. Canada (Commissioner of Patents)* (hereafter *Harvard College*).[7] The federal court trial division, while allowing for patents on the method for introducing the gene and for the preparation of the first generation of mice, rejected the mouse as unpatentable subject matter (Roberts, 1999:36). The federal court of appeals—resting its decision on an interpretation of the mouse as a "composition of matter," consistent with the U.S. Chakrabarty decision—subsequently supported the mouse's patentability (Vaver, 2004:159). Finally, on appeal, the Supreme Court of Canada, in a 5-4 decision, once again rejected higher life forms as appropriate subjects for patenting. The oncomouse was ultimately denied patenting in Canada,

and the Supreme Court decision would seem to have brought some clo-
sure to the issue:

> The Court held that because Parliament chose an exhaustive definition of
> invention, they made an explicit decision to include some subject matter
> as patentable and exclude other subject matter as unpatentable. They ar-
> gued that this exclusion applies to higher life forms. (Atkinson, 2005:15)

Atkinson argues that patenting in Canada followed two general historical approaches. The first approach was an erratic "mixing pot" approach
whereby the requirements of novelty, ingenuity, and utility were blended
with the definition of invention, leading to confusion as to the statutory
definition of "invention" (ibid.:10). The second approach, which gained
prominence in *Harvard College*, assessed whether or not an invention
constituted patentable subject matter, and, if so, independently assessed
whether it had novelty, ingenuity, and utility (ibid.). Thus through *Harvard College* the Supreme Court provided much-needed clarity with respect to determining the patentability of inventions and provided a definitive answer to the question of the patentability of higher life forms:
higher life forms were not patentable subject matter. Nonetheless, the
issue of patentability would arise again in the Percy Schmeiser case.

Prairie Litigation

While technically possible, infringement lawsuits under Canada's Plant
Breeders' Rights Act [PBR Act] have been rare. According to a January
2004 issue of *Germination*, the seed industry finds plant breeders' rights
difficult to enforce (cited in Kuyek, 2005:34). The protection offered by
utility patents is considerably stronger. Their legal requirements would
explicitly preclude farmers' seed saving, even for personal use, in direct
contradiction to the rights afforded by the PBR Act. Post-*Harvard College*, patent protection for seeds had become questionable. Given Monsanto's Technology Use Agreement for its GM seeds, however, even if
patents on seeds were not upheld, farmers who purchased and then saved
Monsanto's seeds would be in contract violation. The TUAs could thus
act to circumvent the potentially "burdensome process of patent infringement suits" (Park, 2003:179), particularly until the uncertain strength of
the patent was resolved.

In contrast to GMO-related lawsuits in the United States, lawsuits in

Canada have been relatively limited. Monsanto itself claims that it has initiated only a handful of cases in Canada, of which only the Schmeiser case proceeded through the courts. Bayer was unavailable for comment. By 2003, Schmeiser was the only utility patent infringement case involving plants in Canada (Kershen, 2004:576, n7). It is not just actual litigation which is at issue, however, as GM detractors claim that the great economic imbalance between farmers and technology developers motivates farmers to accept unfair settlement agreements when faced with an expensive lawsuit. In such cases, settlement can occur before the case is filed, leaving no court record. Given the nondisclosure agreement that accompanies settlement, it is impossible to learn how many such cases there are or how much farmers who settled paid. While the lawyer for the Schmeiser and Saskatchewan Organic Directorate cases states that he has been involved in "a handful" of cases, it is likely that other lawyers across Canada may have had their own "handful" of farmers contacting them for similar preliminary consultation or involvement.

While such statistical data is difficult to come by, there is no shortage of information—or action—surrounding the two court cases that launched the agricultural biotechnology issue onto the Canadian consciousness. With the Schmeiser case, the issue of crop varietal improvements became irrevocably associated with the need for a legal and ethical stance on property and ownership: should farmers' right to save seed be abolished? How should liability for unwanted infringement be resolved? In the Hoffman case, unwanted GM material in organic crops raised further debates over liability and the rights of conventional versus organic farmers. While Schmeiser's case is defensive and Hoffman's is offensive (from the perspective of the farmer), both result from the new practice of granting patents on self-reproducing inventions. These cases alternately put to trial issues such as the validity of patents on life, whether plant patents override plant breeders' rights, the right to farm organically, and the limits of liability for GM patent infringement or for damages arising from unwanted GM material. At heart, these are all questions about the expropriation of farmers' existing rights.

Monsanto v. Schmeiser

The complex web of social and legal ramifications of patented GM seed forced the usually neglected sphere of food production into the public eye when Monsanto Canada, Inc. (hereafter "Monsanto") launched a suit against Saskatchewan canola farmer Percy Schmeiser for patent infringement in 1998. The case was a first not only with respect to its being the first patent infringement suit over GM crops in Canada, but also because Schmeiser alleged that any GM technology in his crops was unwanted. Schmeiser's response to Monsanto's lawsuit was neither passive nor strictly limited to self-defense. Rather, he took on Monsanto offensively. He vehemently fought the suit not only in the law courts but also in the court of public opinion, becoming an icon of resistance to the agricultural biotechnology industry—a "David" versus the "Goliath" Monsanto. The case progressed all the way to the Supreme Court of Canada, which rendered its final decision on May 21, 2004.

The basic facts behind *Monsanto Canada Inc. v. Schmeiser* (hereafter *Schmeiser*)[8] are relatively straightforward, although much occurs below their surface. Percy Schmeiser was a conventional canola farmer from Bruno, Saskatchewan. He was one of the roughly 20% of canola farmers in western Canada estimated at trial to still practice seed saving.[9] When Monsanto introduced its Roundup Ready canola in 1996, Schmeiser did not adopt it and continued his routine practice of saving and cleaning a portion of his canola seed for planting the following year. He believes this practice helped him select for his own strain of canola, which was more disease resistant: between 1993 and 1999 he purchased no new canola seed.[10]

Schmeiser stated at trial that he subscribes to minimal chemical use and prefers chemicals that can be incorporated into the soil, rather than Roundup, as he believes they reduce moisture loss. However, Roundup was a tool he used, as did many other farmers, to burn off his fields prior to planting and to keep the area in ditches and around power poles clear. In 1997, Schmeiser conducted such a routine spraying of ditches and power poles. Unlike in the past, however, the volunteer canola he sprayed did not die with the weeds. Schmeiser decided to conduct a test, and sprayed approximately three acres worth of Roundup on a test strip of his canola crop, parallel to the power poles. He found after several days that a large portion of it—he estimates about 60%—did not die. Despite this knowledge, Schmeiser stated that he continued his routine practice of cleaning and saving seed, and replanting it for the 1998 crop.

Purportedly acting on a tip that Schmeiser was farming their RR canola without licence, Monsanto approached Schmeiser and ultimately sued him for patent infringement in 1998. According to Schmeiser, it hit him like a lightning bolt when he was served with Monsanto's claim that he had infringed their patent on the canola:

Well, I had no idea or nothing what was, you know, that I had been always using my own seed, I never bought any Monsanto seed and all of a sudden I'm getting a Statement of Claim that I was using their product.[11]

The details of *Schmeiser* are more fulsome and contested than can be outlined here, though a few points bear mention. Monsanto, for example, originally accused Schmeiser of "brown bagging," a term that applies to the illegal sale of certified or patented seed between farmers (hence, transferred in unmarked bags). Despite efforts to support this claim through interviews and other investigative means, however, Monsanto dropped this allegation at trial. It is further significant that Monsanto also dropped its claim of infringement with respect to Schmeiser's 1997 crop, and proceeded only on the basis of his 1998 crop, and the allegation that he had knowingly and deliberately segregated, saved, and then propagated their patented RR canola in this crop. Monsanto pointed to the fact that, on the basis of its own tests, the canola from samples of Schmeiser's 1998 crop was 95% to 98% Roundup tolerant.[12]

In response, Schmeiser claims to have been a victim of GM contamination. He testified that after his Roundup spray test, the approximately 60% of canola that remained was thickest closest to the road, thinning as it progressed into the field, suggesting that the GM material might have spread from there. He vehemently denied Monsanto's conclusions regarding the percentage of GM canola in his crops, although supporting evidence is weak, and raised significant objections about the sampling procedure: samples were collected by Monsanto without independent oversight; discrepancies arose around the location and transfer of these samples; and evidence for the presence of the patented gene was based on tests conducted by Monsanto's own staff.

As the Supreme Court of Canada accepts the facts of the case, the subject of the complaint is traced to Schmeiser's field no. 1, on which Schmeiser sprayed his Roundup test. In the fall, this patch of surviving canola was harvested, separated, and stored over the winter. Despite being notified by Monsanto that it had concerns he was growing their RR canola, Schmeiser then treated and planted these seeds in nine other fields in the fall of 1998.[13] While these are the facts as determined by the court, the chronology of how Schmeiser might have acquired a full crop of RR canola, if one accepts that this is indeed the case, is a complex and disputed record of treated and untreated seed from different fields, deposited in various field granaries, recombined, and used to reseed. Nonetheless, the most salient aspect relates to the field where Schmeiser conducted his test, resulting in an area with a high concentration of RR canola.

Schmeiser testified that he himself did not combine or swath the field in question as he had an injury at the time, and he had his neighbor, a hired hand, do it. This hand was provided with no special instructions, although Schmeiser noted that given the practicalities of navigating the farm equipment, he usually approached the field from a particular way. The first load of canola was dumped into an extra truck parked in the field, as is the normal procedure when the primary truck doesn't return from the granary in time. This truck later failed to start, however, and it was tarped and left in the field, and a few months later was moved to Schmeiser's storage quonset in Bruno. Schmeiser claimed that when he wanted to seed the following year, the canola in this truck was the quickest to access, as it would not be hampered by bans restricting the maximum load that could be hauled on municipal roads in springtime. Consequently, it was this canola that he brought to be cleaned and subsequently seeded his crop with in 1998.

However such a chronology is viewed, in March 1998 Monsanto sent Mr. Robinson, from Robinson Investigations, to talk to Schmeiser regarding their suspicions about his canola. Schmeiser claims to have gotten angry at Mr. Robinson's statement that they had sampled his fields, but when Mr. Robinson left, Schmeiser reportedly forgot about the incident. Nonetheless, the fact of this interaction, if not the spray test itself, meant that Schmeiser was aware of the presence of Monsanto's RR canola in his fields.

Although the original practical question for the court case concerned whether or not Schmeiser deliberately infringed on Monsanto's patent, this practicality was almost immediately overshadowed by the case's broader significance for farmers' rights. While responding to the specifics of the claim against him, Schmeiser effectively put a number of the technologies' unchallenged aspects on trial. Schmeiser argued that he had never deliberately planted, or caused to be planted, any patented RR seeds. He claimed that given the gene's "unconfined release" into the environment, the plaintiffs had not controlled its spread, and thus "lost or waived their right to exercise an exclusive patent over [it]."[14] Further, he argued that the patent was invalid and void because a self-propagating life form was not the proper subject matter for a patent. He also claimed that a finding of infringement would constitute granting a patent on a plant, and that this was not possible in Canada in light of the Plant Breeders'

Rights Act. A further question was the scope of the patent: given that Schmeiser claimed not to have sprayed Roundup on his crop (excepting his test strip), he claimed not to have "used" the invention, and thus could not have infringed the patent.

Federal court trial judge MacKay took a purposive approach to the claims of the patent, asserting that any interpretation should be "fair and reasonable to both the patentee and the public."[15] With respect to the argument that the Plant Breeders' Rights Act represented Parliament's intent that plants should be governed by legislation other than the Patent Act, the judge found nothing in it that "precludes an inventor from seeking registration under the Patent Act."[16] This decision in support of patenting plants thus supports a prohibition on seed saving (otherwise permissible under the PBR Act). Further, while acknowledging that replication of the gene might occur in nature, Judge MacKay found that this did not compromise the patent. While not all progeny of the RR plants will be Roundup tolerant, "those plants containing the gene can be subject to Monsanto's claims as patent holder."[17] The implication was that the patent will stand through as many generations of the gene as are produced within the time period of Monsanto's patent. This is a far simpler dispensation of the debate over patenting subsequent generations than we shall see in Mississippi.

Lastly, the judge considered the validity of the patent in the context of the Harvard College case regarding the oncomouse. At the time of the Schmeiser trial decision, *Harvard College* had only reached the federal court of appeals, and its ruling that the Patent Act did not exclude a patent on an appropriate case regarding a nonhuman higher life form. Judge MacKay ruled that while *Harvard College* was not directly applicable, as it related to a higher life form, it implicitly supported Monsanto's patent. Further, he noted that in *Abitibi* the Patent Office supported that "certain life forms may be patentable."[18] This consideration of where plants stood in relation to higher life forms would come up again.

As noted, significant flags were raised over sample collecting and genetic testing. For example, as is usual in the course of their business, Humbolt Flour Mills took a sample of Schmeiser's 1998 crop when he brought it in to be cleaned. They then provided this sample to Monsanto without Schmeiser's knowledge or consent. The judge declined to rule on the propriety of the evidence collection and noted that Schmeiser had civil remedies to address such issues. Further, while acknowledging that the court needed to weigh the evidence from the genetic tests carefully, there was insufficient cause to disregard it. Overall, despite the concerns over evidence sampling and handling, the judge reached the "tentative conclusion" based on the "balance of probabilities" that the plaintiff had infringed the patent.[19]

In sum, Judge MacKay found no justification to rule against the validity of the patent and found that Schmeiser had infringed it. With respect to Schmeiser's defense that there was no intention to infringe the patent, the judge ruled that this was immaterial: "it is well settled that infringement is any act which interferes with the full enjoyment of the monopoly rights of the patentee."[20] While it was the case that the "invention has utility in resistance to glyphosate," the judge emphasized that "none of the [patent] claims specifies this utility nor does it require the use of glyphosate."[21] Consequently, in the new context of self-reproducing inventions, the trial court decision suggests that an "act of interference" simply entails knowledgeable possession.

Significantly, while the source of Schmeiser's GM canola was not determined at trial, the defense produced a number of possible points for contamination, including a neighbor whose tarp was loose as he hauled GM canola past Schmeiser's fields, and another whose swaths blew onto Schmeiser's land. Five canola growers grew RR canola in Schmeiser's area.[22] Nonetheless, the cause of the RR gene's presence was ultimately deemed irrelevant. Judge MacKay ruled that the source of the RR canola was "really not significant for the resolution of the issue of infringement."[23] While this determination was based on the fact that Schmeiser "knew or ought to have known" that his seeds were Roundup tolerant, it soon became one of the most notorious statements of the trial. Schmeiser himself credits this statement with fueling resistance to the changes that were occurring in agriculture:

> After [the trial], then alarm bells I think went off around the world in how farmers, and organic farmers, could lose their rights to their seeds and plants overnight through contamination or pollution against their wishes. Overnight, can no longer use their seeds or plants. And I think, if the judge wouldn't have made that decision, or had worded it somehow else, it would never have become an international issue. That wording is what did it. (K#2, Schmeiser)

The trial judge's ruling appeared to interpret events strictly from the perspective of patent law, with little leeway for the special case of inventions that were self-reproducing, and despite there being no denying the prolific nature of canola or the presence of GM canola volunteers just a few years after its introduction. The likelihood of farmers facing un-explained and unwanted presence of GM material was not in dispute. In fact, the trial judge heard testimony from two Saskatchewan farmers about their experience with unwanted RR canola volunteering in their fields, despite neither having previously grown the crop. Louis Gerwing, for example, testified about his problem in 1999:

> On one quarter I had chemfallow. I had sprayed it twice with Roundup and these [canola] plants never died, they were growing along the high-way and out into the field quite a ways. . . . Well, most of it was along the highway and then there was several plants I would say a thousand feet down the field, just about a quarter mile some were out.[24]

Similarly, Charles Boser testified about his experience when he at-tempted to chemfallow his field that same year. He had a custom appli-cator spray the field twice to combat weeds, after which everything was "burned to a crisp . . . except for the canola."[25] He indicated that well over 100 out of 160 acres had RR canola on it, "in a pattern that went com-pletely across the whole quarter section, from east to west."[26] According to trial testimony, Monsanto had attended to a handful of similar com-plaints from other farmers.

Monsanto did not deny the farmers' testimony. In fact, they used the cross-examination to expound on their willingness to remove any un-wanted RR canola reported to them—as they did for both these farmers—thus demonstrating their continued control over their invention. Mon-santo's conduct with respect to these farmers, in conjunction with the Technology Use Agreement, was indeed taken by Judge MacKay as evi-dence of the company's effort to control the spread of its invention, and supported his conclusion that there was no loss or waiver of Monsanto's exclusive rights over it. The judge did not specifically discuss the signifi-cance of the increasing natural spread of the gene, if any. Therefore he left ambiguous whether there was any amount of natural spread that could constitute a loss of control, as long as Monsanto continued to demon-strate an explicit intention to control it. What is unambiguous about the ruling, however, is Judge MacKay's determination of the proper purview of patent rights and property rights:

> For the defendants it is urged Monsanto has no property interest in its gene, only intellectual property rights. While I acknowledge that the

seed or plant containing the plaintiff's patented gene and cell may be owned in a legal sense by the farmer who has acquired the seed or plant, that "owner's" interest in the seed or plant is subject to the plaintiffs' patent rights, including the exclusive right to use or sell its gene or cell, and they alone may license others to use the invention.[27]

Thus a farmer whose field contains seed or plants originating from seed spilled into them, or blown as seed, in swaths from a neighbour's land or even growing from germination by pollen carried into his field from elsewhere by insects, birds, or by the wind, may own the seed or plants on his land even if he did not set about to plant them. He does not, however, own the right to the use of the patented gene, or of the seed or plant containing the patented gene or cell.[28]

The findings of the trial judge thus render an explicit determination that patent rights trump the property rights of farmers. This leaves farmers potentially liable for any patented GM material in their crops, regardless of how it got there and whether they exploited its benefits, as long as they were aware of its existence.

While Schmeiser was liable because he "knew or ought to have known" that his crop was Roundup tolerant, the ruling left undetermined the status of farmers who were not aware of GM presence in their crops. In response to concerns that a finding of infringement would prompt farmers to curtail seed saving out of fears of liability for contamination, the judge emphasized that Schmeiser's contamination was "not simply from occasional or limited contamination."[29] He did not, however, suggest how such a scenario would be viewed except to note that Monsanto assisted with the removal of canola by those farmers who contacted the company. He left further undetermined what would be required to establish a claim of ignorance of contamination, and what the outcome would be for farmers who were ignorant of it.

The decision of the trial judge seemed to confirm the worst fears of biotechnology's watchdogs. *Schmeiser* brought to light the nontechnological changes that biotechnology brought to farming, and raised the prospect that ordinary farmers, farming in the same manner they always had, could find themselves liable for patent infringement. While the contractual conditions associated with the use of the technology raised the ire of some who wished to use the technology, maintaining patent rights over the involuntary presence of it meant that something could be taken away from even those farmers who did not ultimately consent to it. Consequently, the *Schmeiser* decision raised huge concerns that biotechnology represented a transfer of control over agriculture to corporate hands: expropriationism by act of nature.

Having had their patent affirmed, both Monsanto Canada and the Monsanto Company (U.S.) sought remedies for Schmeiser's infringement: the former sought $15,450.00 for the unpaid technology fee of $15/acre applied to Schmeiser's 1998 crop; and the latter sought profits of $105,000.00 based on Schmeiser's accounting of revenues and costs of production. In addition they sought exemplary damages. Judge MacKay rejected the claim for dual remedies, exemplary damages, and personal liability. He also rejected the conclusion that the $105,000.00 represented Schmeiser's profits, as it provided no deduction for his labor. Rather, MacKay indicated that the plaintiffs could jointly claim for profits (to be determined) or for general damages in the amount of $15,450.00. The plaintiffs opted for profits, and were awarded $19,832 plus prejudgment interest, postjudgment interest, and costs.[30] The judge also granted an injunction preventing Schmeiser from planting any seed retained from his 1997 or 1998 crop, or for which he knew or ought to know was Roundup tolerant, and he was ordered to relinquish any plants or seeds saved from these years.

Given the radical shift in farmers' rights that Judge MacKay's decision signaled, and the existence of numerous environmental and civil society groups already involved in the GM issue, *Schmeiser* was widely publicized. The case became a lightning rod for those opposed to patents on life; increased corporate power; and the social, health, and environmental risks of biotechnologies. Schmeiser launched a website for his cause and became a celebrated public speaker and anti-GM activist. He also appealed his case.

Many of Schmeiser's 17 points for appeal replicated his claims from the federal court, such as issues related to sampling and remedies. The appeals court upheld the trial court with respect to these matters. Monsanto also cross-appealed, predominantly with respect to increasing damages, but was denied on the grounds that an accounting of profits was an equitable remedy, and that the trial judge had not erred in the award. With respect to substantive issues, Schmeiser's defense focused on claims aimed at defining the scope of patent rights for a self-reproducing invention which challenges the property rights of farmers: how broadly can such a patent be construed and what is the status of those who come to inadvertently possess it?

The defense argued that the patent was too broadly construed if it allowed for infringement where a RR canola plant was not sprayed with Roundup. It contended that such an interpretation was "unfair" to the public, "because if it stands, Mr. Schmeiser could find himself liable for infringement simply by following his normal farming practices."[31] The appeals court again responded through a purposive construction of the patent itself, establishing whether there had been any interference with its "full enjoyment." It found that the essence of the claim was the presence of the gene, and spraying Roundup was immaterial. Second, the defense argued that the source of the RR canola was not irrelevant, and that Schmeiser could not have infringed because he "took no steps to cause glyphosate resistant canola plants to grow on or adjacent to his property in 1997."[32] To burden his crop with a patent claim was thus "an unjustified intrusion on Mr. Schmeiser's property rights."[33] The court found that as the infringement claim only pertained to Schmeiser's 1998 crop, and that as its source of GMOs was determined to be the result of him saving seed that he knew or ought to have known was RR, then the original source was irrelevant. The court responded to the remaining claims under three headings: conflict of rights, innocent infringer, and the effect of unconfined release.

The court of appeals found that the issue of a conflict between property and patent rights is usually only relevant when establishing remedies, but that where there is a conflict, "the jurisprudence presents a number of examples in which the rights of ownership of property are compromised to the extent required to protect the patent holder's statutory monopoly."[34] That is, the rights of patent holders subvert the property rights of farmers. The court did acknowledge that accidental infringement can occur, however, and it found merit in the defendant's claim that support of the patent in *Schmeiser* could disadvantage any farmer who didn't grow RR canola, putting such farmers at risk of infringing Monsanto's patent:

> There is considerable force to the argument that it would be unfair to grant Monsanto a remedy for infringement where volunteer Roundup Ready Canola grows in a farmer's field but its resistance to glyphosate remains unknown, or if that characteristic becomes apparent but the seeds of the volunteer plants are not retained for cultivation.[35]

Consequently, while intention is generally considered immaterial to infringement, the appeals court found that it "seems . . . arguable that the patented Monsanto gene falls into a novel category."[36] On this basis, the court found it to be "an open question" whether in circumstances where a farmer was either unaware of the patented gene or was aware but did nothing to promote its presence, Monsanto could "obtain a remedy for infringement on the basis that the intention of the alleged infringer is irrelevant."[37] The court found that Schmeiser's propagation of plants that he "knew or ought to have known" were Roundup tolerant set him apart from this category, and consequently the decision in *Schmeiser* need not resolve the relevance of intention in the above scenarios.

Last, with respect to whether the unconfined release of RR canola into the environment—and its subsequent spread—was tantamount to a waiver of the rights of the patent holder, the defense argued that it was "physically impossible" to prevent the spread of RR canola and that whatever steps Monsanto had taken to prevent its spread were "curative rather than preventative."[38] The court of appeals nonetheless upheld the findings of the trial court. It found that even if the defense was correct on these points, it would not amount to a waiver, but might "cause Monsanto some difficulty in defending its patent rights in certain situations."[39]

While the appeals court dismissed the appeal, it more explicitly defined some limitations to a finding of infringement as it might apply to those the court considers truly "innocent" infringers. For example, Schmeiser protested the injunction that prevented him from saving canola seeds that he knew or ought to have known were Roundup tolerant; given the prolific nature of canola, he could "reasonably anticipate the constant presence of volunteer glyphosate resistant canola in his field at all times."[40] Consequently, he argued that the injunction prohibited his normal practice of seed saving and forced him to purchase new seed every year. The appeals court did not accept this interpretation of the injunction, however, which it stated did not refer to the "awareness of every Canadian farmer" about the possibility of RR volunteers. Rather, the "requisite knowledge would not be established unless Mr. Schmeiser, because of the use of Roundup or some means of chemical testing, knows or is wilfully blind to the presence of glyphosate resistant canola plants."[41] Thus as long as Schmeiser didn't test the seed before he saved it, and conducted no other questionable actions, he would not violate the injunction. Following this logic, it seems likely that this court would also find that possession without testing would not constitute infringement for another who was similarly "innocent" of knowledge. Were this perspective on innocence to be maintained, it at least provided some clarity to the legal quagmire raised by granting patents on self-reproducing, visually indistinguishable, GM canola.

Considering the great personal and financial cost of litigation, it would have been understandable if Schmeiser had dropped his fight at this point. The decisions were not in his favor, and the judgment against him was just under $20,000—relatively small considering the gamble involved in proceeding. Schmeiser did not stop, however, but applied—and was accepted—for his case to be heard by the Supreme Court of Canada.

The issues under consideration by the Supreme Court again primarily related to the scope and validity of the patent, and to what remedies Monsanto was entitled if Schmeiser was found to have infringed. The applicability of the patent to plants, and the meaning of "use" were key elements of this decision. The Supreme Court returned a 5-4 split decision, upholding the patent, with the minority dissenting in part. Notably, the Supreme Court explicitly stated that its role was not to consider the "wisdom or social utility" of GMOs. Perhaps even more telling, the court also stated that "the innocent discovery of farmers of 'blow-by' patented plants on their land or in their cultivated fields" was not under consideration. Rather, the court's decision was solely concerned with "the application of established principles of patent law to the essentially undisputed facts of this case."[42] Schmeiser's knowledge of the presence of the Roundup-tolerant technology in his canola was again an important factor.

The majority found Monsanto's patent to be valid, as Monsanto's claims did not extend to the plant itself, which would be unpatentable, but were for the genes and the modified cells that made up the plant. It also found that "whether or not patent protection for the gene and the cell extends to activities involving the plant is not relevant to the patent's validity."[43] A purposive construction of the invention recognized that it would be practiced in plants, and this was sufficient. A purposive approach was further evident in considering whether the defendant "used" the patent. The majority determined that whether the patent was "used" depended on whether the defendant actively "deprived the inventor . . . of full enjoyment of the monopoly conferred by law."[44] This can occur through people employing the invention to their advantage—even if they did not actually utilize its herbicide-tolerant properties—such as through the "stand-in utility" of the invention. For example, even though Schmeiser did not spray Roundup, he could have decided to do so in the future or he could have sold the technology to other farmers. This assessment of "use" is thus constructed solely from the perspective of the patentee, and what the patentee would lose with respect to its patent rights; there is no equivalent incorporation of the public's gains or losses. This construction of the patent in such a narrow way that it neglects the concept of fairness became a point of contention for the minority.

The court also considered whether Schmeiser infringed the patent by using the plant, when it was the gene and cell that were patented. According to the majority, infringement occurs when the defendant uses a patented part even if it is contained in something unpatentable. They likened this to patented Lego blocks assembled in an unpatented structure. Using this conception, Schmeiser's claim not to have used the invention by cultivating plants "flies in the face of century-old patent law, which holds that where a defendant's commercial or business activity involves a thing of which a patented part is a significant or important component, infringement is established."[45] According to the majority, infringement of Monsanto's patent does not require the "use of the gene or cell in isolation."[46] As we shall see, the minority strongly disagreed with this interpretation.

While intention was irrelevant for infringement, the court found that its absence "may be relevant to rebutting the presumption of use (and thus infringement) raised by possession."[47] Therefore, possession constitutes infringement, but where the defendant makes no effort to gain advantage from the invention and "can show that the object is held without a view to advancing the defendant's interest," it may provide a defense.[48] Once again, the court was clear that it did not consider Schmeiser to be an "innocent bystander," based on the fact that Schmeiser had saved and planted seeds—whatever their original source—that he knew or ought to have known were Roundup tolerant. He could have rebutted the "presumption of use" by trying to arrange for the technology's removal, otherwise showing that the presence was accidental or unwelcome, or having a concentration in his fields consistent with "blow-by" canola. In a scenario that smacked of active cultivation, the court considered the presumption of use unrebutted:

> Knowledge of infringement is never a necessary component of infringement. However, a defendant's conduct on becoming aware of the presence of the patented invention may assist in rebutting the presumption of use arising from possession.[49]

In this way, the majority clearly expresses the difference with respect to their decision in *Schmeiser* from that they might make in a case involving what they considered an innocent infringer. Nevertheless, there are no such qualifications with respect to how they construed the issue of effective (if not technically legal) patents on plants, or with respect to whose rights prevail in any conflict between patent and ownership rights. With respect to the defendant's claims that the "ancient common law rights of farmers to keep that which comes onto their land" were being violated, the court's response was straightforward: "[T]he issue is not property rights, but patent protection. Ownership is no defense to a breach of the *Patent Act*."[50]

The Supreme Court decision was not unanimous, however. Most significantly, the minority, led by Justice Arbour, found far greater importance in the fact that the appeals court's decision on *Schmeiser* came prior to the Supreme Court's final decision in *Harvard College*. While *Harvard College* allowed claims for the process of producing a higher life form (as long as it required significant technical intervention by man) and for lower life forms, it excluded the patentability of higher life forms. Consequently, the minority in *Schmeiser* felt that the central issue was whether the appeals court's decision could stand in light of this exclusion. The majority had dispensed with this question rather quickly on the basis that the cases differed because the patent refused in *Harvard College* was for a mammal, and because the patent commissioner had already allowed claims for plasmid and somatic cell culture.[51]

The minority also countered the majority's emphasis on the commercial interests of the patentee as the essential elements of the patent. Rather, Justice Arbour argued that there are further issues involved in a purposive construction of a patent. For one, the scope of the patent must be such that whatever is not explicitly claimed is considered disclaimed — "an inventor cannot enlarge the scope of the grant of exclusive rights beyond that which has been specified."[52] This is significant, as the specifications of Monsanto's patent clearly do not claim the plant. From this perspective, Monsanto's claims are valid as long as they are not construed as extending beyond the cell and the gene to the plant. According to Justice Arbour: "In order to avoid the claim extending to the whole plant, the plant cell claim cannot extend past the point where the genetically engineered cell begins to multiply and differentiate into plant tissues, at which point the claim would be for every cell in the plant, i.e. for the plant itself."[53]

A further, related, issue for a purposive construction of the patent is that claims should be fair and reasonably predictable for the public. Plants are unpatentable, and consequently "[t]he public is entitled to rely on the reasonable expectation that unpatentable subject matter falls outside the scope of patent protection and its use does not constitute an infringement."[54] This concern with fairness also includes the hypothetical person "skilled in the art" that is making and interpreting patent claims. According to the minority, such a person "could not reasonably have expected that [Monsanto's] exclusive rights for gene, cell, vector, and method claims extended exclusive rights over unpatentable plants and their off-spring."[55] As Monsanto's patent claims could not extend to whole plants, the minority concluded that there could be no infringement, and the majority had erred in its reading of the patent:

> [T]he test for determining "use" is not whether the alleged user has de-prived the patentee of the commercial benefits flowing from his inven-tion, but whether the alleged user has deprived the patentee of his mo-nopoly over the use of the invention as construed in the claims. . . . the lower courts erred not only in construing the claims to extend to plants and seeds, but also in construing "use" to include the use of subject-matter disclaimed by the patentee, namely the plant.[56]

Once Schmeiser's case reached the Supreme Court, many of the orga-nizations that were interested in it had the opportunity to expand beyond their public canvasing role by applying for intervener status, which would afford their perspectives a formal hearing. Eleven groups were ultimately granted this status. Four of these were supportive of Monsanto's right to patent seeds: the Canadian Canola Growers Association [CCGA], Ag-West Biotech, Inc., the Canadian Seed Trade Association, and BIO-TECanada. The latter three are seed trade and biotechnology industry groups. They were concerned that if plant technology developers were unable to protect their intellectual property, technological development would be negatively affected, to the detriment of both the Canadian bio-technology industry and the seed sector.

Considering the issue of litigation against potentially innocent farmers, the support of the CCGA (an umbrella organization for provincial canola grower organizations) for Monsanto's claim might seem surprising. Its position was that while it is not very common for canola producers to save seed, they do export a large portion of their canola, and any com-promise of their access to GM technology could affect their global com-petitiveness. Consequently, the CCGA argued that intellectual property protection was necessary to encourage research and maintain access to new developments. In order to advance their position, the CCGA took on the issue of the legitimacy of patenting directly, arguing that plant genes, cells, and plants are, in fact, the proper subject for patents as they are not higher life forms: they "are not sentient, do not express emotion and are mere 'compositions of matter.'"[57] This argument reveals the ambiguity inherent in even the definition of "higher life form." In short, the CCGA's concerns over losing the technology were far greater than over any potential liability risk for farmers:

> Farmers' privilege to save seed and potential innocent infringement of patents are manageable issues given current agronomic practices. No remedy would be awarded to a patent holder who sued an innocent bystander in a frivolous infringement claim. If this issue becomes a concern to the CCGA members, the CCGA will propose to Parliament that it amend the Patent Act to provide regulatory protection for farmers.[58]

On the opposing side, 7 of the 11 organizations granted intervener status were supportive of many of the issues Schmeiser raised. Six of these submitted jointly: the National Farmers Union; the Council of Canadians; the Sierra Club of Canada; the International Centre for Technology Assessment; the Research Foundation for Science, Technology, and Ecology; and the Action Group on Erosion, Technology, and Concentration. These groups were concerned with social issues—such as the broader impact of patents on life and seed-saving restrictions—that had not found a forum in government regulation or in the lower courts, whose rulings, they argued, failed to consider the public interest:

> An overly broad interpretation of patent claims may not only interfere with further innovation, a traditional concern of patent law, but also derogate from the existing rights of third parties and adversely affect the environment and biodiversity. This significantly complicates the task of finding the proper balance between public and private interests when patents concern living organisms which spread and interact with the environment.[59]

Unfortunately for these groups, the majority of the Supreme Court was resolute in not considering any issues outside the Patent Act. As in *Harvard College*, the court determined that if these issues required consideration, it was up to Parliament to do it:

> Inventions in the field of agriculture may give rise to concerns not raised in other fields—moral concerns about whether it is right to manipulate genes in order to obtain better weed control or higher yields. It is open to Parliament to consider these concerns and amend the *Patent Act* should it find them persuasive.[60]

Further: "Where Parliament has not seen fit to distinguish between inventions concerning plants and other inventions, neither should the courts."[61]

The scope of the patent still falls within this very limited concern, however, and it is on this point that the attorney general of Ontario intervened, advocating that the court "apply its usual approach of interpreting patents narrowly to gene and DNA sequence claims."[62] The attorney general's specific interest concerned health care, and that the Schmeiser decision "not inadvertently restrict the ability of researchers and health care providers . . . to develop new tests and treatments for patients."[63] This concern resonates with the minority's position on determining the scope of the patent. Nonetheless, the majority accepted the much broader scope.

The Supreme Court's only significant divergence from earlier court rulings was in regard to remedies. As noted, Monsanto had chosen an accounting based on profits, which for Schmeiser's 1998 crop was calculated at just under $20,000. The Supreme Court ruled that the inventor is only entitled to the portion of the profit that can be attributed to the invention, however. Schmeiser sold his RR canola for the same amount as he would have any canola, and, as he did not spray Roundup, his yields and dockage also did not differ. Consequently, the invention had not contributed anything to Schmeiser's profits, and Monsanto was awarded no remedies. The court further declared that on the basis of the mixed results, both sides would have to pay their own costs. This decision on remedies could act as a safeguard for anyone in the "innocent infringer" predicament: as long as no Roundup was sprayed, and the seed was marketed in the normal channels, Monsanto would get no accounting of profits. This would have been the end of the innocent infringer concern if it weren't for the fact that Monsanto could still elect for a remedy based on damages—at $15/acre plus any further proven damages—instead of profits. Further, while Monsanto may gain only modest compensation from such litigation, it still holds a highly punitive aspect for farmers, given it is costly, time-consuming, and high risk. Consequently, remedies notwithstanding, Monsanto retains a strong motivation to sue in order to maintain control of its invention.

On The Offensive: *Hoffman v. Monsanto*

In *Hoffman v. Monsanto Canada Inc.*[64] (hereafter *Hoffman*), organic farmers Larry Hoffman and Dale Beaudoin filed for class action certification in a claim against Monsanto and Aventis (subsequently Bayer) over the loss of their organic canola market. *Hoffman* was launched in 2002. Certification was denied in the Saskatchewan court of queen's bench in May 2005, and again in the Saskatchewan court of appeals in May 2007. An application for leave to appeal at the Supreme Court of Canada was denied in December 2007. While *Hoffman* ostensibly died in its legal infancy, it nonetheless garnered significant debate over the issues involved, as it was the first case in either Canada or the United States where agricultural producers offensively attempted to assign liability to biotechnology developers for the impact of their technology.[65]

While Larry Hoffman and Dale Beaudoin were the representative plaintiffs for the legal action, it was spearheaded by the Saskatchewan Organic Directorate [SOD], an umbrella organization for organic production and related enterprise in Saskatchewan. This fact would later cause some complications for the claim. The SOD was concerned over the introduction of GM crops into the environment because organic producers are not allowed to use any synthetic pesticides, fertilizers, or GM organisms, as a condition of their organic certification. With the prolific nature of canola, however, organic farmers risk GMO transfer through the same involuntary means (e.g., wind, combine, trucks) raised in *Schmeiser*, albeit with different consequences. They are also at risk from the general infiltration of GMOs in the seed supply as "few, if any, pedigreed seed growers in Saskatchewan will warrant their canola seed to be GMO-free."[66] According to the *Hoffman* statement of claim: "Contamination of organic products by prohibited substances such as GMOs can result in the rejection of shipments and substantial losses to organic farmers."[67] The production of organic canola essentially ceased, and organic producers had to remove canola from their crop rotations as a result of their inability to keep it GMO-free. The individual claims of Hoffman and Beaudoin supported this contention.

Both Larry Hoffman and Dale Beaudoin are certified organic through the Organic Crop Improvement Association [OCIA] International. Larry Hoffman farmed transitionally beginning in 1989 and gained certification in 1991. He farmed approximately 2,400 acres in Spalding, Saskatchewan, first growing organic canola in 1994. Under cross-examination, Hoffman stated that he would have grown canola again in 1997, in a normal rotation, but decided against it. While he wasn't aware of any GM contamination in his fields, he had become increasingly sensitive to the possibility of it, and didn't want to risk his crop being "rejected or put on hold."[68] Hoffman stated that given the evidence of contamination around him, "I would assume that if I have a crop of GMO canola beside me . . . that I'm going to get drifted too."[69] Instead of growing canola, therefore, Hoffman grew rye, peas, and barley, calculating that this decision cost him somewhere between $22 and $100 per acre, for 181 acres, depending on the price he might have received for canola.[70]

Dale Beaudoin farmed approximately 600 acres in Mayfield, Saskatchewan. He began farming organically in 1988, but was only certified in 1995. His main crops were wheat, oats, canola, and barley, and he grew canola every year from 1995 to 1999.[71] In 1999, Beaudoin signed a lucrative contract to produce organic canola for $16.50 a bushel.[72] The contract required end-product testing, with a GMO tolerance threshold of 0.01%. Despite following all OCIA production guidelines, however, Beaudoin's crop did not meet the threshold, and he lost his contract. He subsequently managed to sell his crop through regular organic channels, albeit at the far lower rate of $13.25 per bushel.[73] As a result of this incident, and the fear that it could occur again and ultimately exclude him from even these regular organic channels, Beaudoin ceased growing canola.

The SOD was not simply concerned about the loss of canola, however; RR wheat was on the horizon. Given that GM canola had effectively put an end to the fledgling organic canola industry, the prospective introduction of RR wheat was a considerable threat, and a prime motivation of the class action was to try to stop it. Consequently, the action not only sought redress for damage to the canola industry, but also an injunction against the release of RR wheat. When Monsanto withdrew its RR wheat application in June 2004, it was a huge victory for organic farmers, but a double-edged sword for their class action, for which wheat provided a much stronger case. Wheat was a pervasive crop that could draw many Saskatchewan organic farmers into their "class" for action, whereas canola was much more selectively produced. Further, organic certification standards prohibiting GMOs were only in place after GM canola's introduction, but prior to RR wheat's. Given the SOD's concerns about future GM introductions, they decided to proceed nonetheless.

In 2001, the SOD set up a committee in order to facilitate the lawsuit—the Organic Agriculture Protection Fund—the function of which was to raise funds, handle media, publicize their cause through a website, and manage the affairs that surrounded the litigation. The Class Actions Act was on the horizon in Saskatchewan, and the plaintiffs delayed their action until it was brought into force, on January 1, 2002. The *Hoffman* statement of claim was submitted on January 10, 2002. As in *Schmeiser*, the significance of *Hoffman* extends far beyond the immediate facts of the case. While the *Hoffman* claims relate to damage to the organic canola industry, the case has two very significant features, one content and one format related.

First, the subject matter of *Hoffman* is unique in that it represents an attempt by organic farmers to make technology developers responsible for any negative impacts of their technology. Organic farmers were not simply concerned about the loss of crops in their rotations, although this was certainly a significant issue, but about what that loss meant more broadly. They argued that no less was at stake than "the right to grow organic crops; the right to serve organic markets; the right to eat GMO-free food; and the right to farm organically" (SOD, n.d.). Essentially, they argued for their right to provide an alternative to industrial agriculture, and against any infringement on this right. In essence, this is an argument against the ultimate act of expropriation: ending the organic industry and converting organic lands into industrial (GM) crop lands.

Hoffman went on the offensive against this act of expropriation. The claim asserted that when Monsanto and Aventis released their GM canola they "knew or ought to have known" that it would contaminate the environment and damage organic production:

> The Plaintiffs state that the Defendants knew, or ought to have known, that the introduction of GM canola into the Saskatchewan environment without any, or in the alternative, proper, safeguards would result in GM canola infiltrating and contaminating the environment, seed supplies, and property of certified organic grain growers.[74]

In this manner, the plaintiffs attempted to hold the biotechnology industry to the same requirements of awareness, attention, and action regarding the self-replicating technology that was implicitly being imposed on farmers in *Schmeiser*. If *Hoffman* were to be successful in establishing corporate liability in this way, it would represent a phenomenal social shift, forcing industry to be attentive to both the costs and the benefits of retaining ownership over such self-reproducing inventions.

The second significant feature regards format. As demonstrated in *Schmeiser*, the economic and power imbalance in any dispute between farmers and technology developers raises significant social concerns given the high personal and financial cost to farmers—regardless of the legal outcome. This imbalance was furthered by whose rights took precedence in the clash between property and patent rights. Consequently, there are significant concerns that technology developers have an unfair advantage in litigation. If organic farmers could proceed through a class action, however, it would significantly mitigate some of their disadvantage. First, the costs would not be borne by an individual, but through a fund supported by organic farmers more broadly, and the cost of multiple lawsuits would be avoided. Second, by filing under the Class Actions Act, the plaintiffs would not be responsible for the defendants' costs in the event they failed, a significant factor for expensive litigation. Perhaps most importantly, in the event of a legal win, any resulting award would be multiplied for the whole of the class, and so would have a far greater impact on the defendants, potentially even curtailing future GM introductions.

Given the plaintiff's preferred format, the first step was to gain class action certification, which required a number of criteria. The plaintiff had to demonstrate that (1) it had a cause of action, (2) it had an identifiable class, (3) it had a class with common issues, (4) class action was the preferable procedure, and (5) there was a representative plaintiff.[75] The evidence submitted for class action certification is only preliminary; its purpose is to demonstrate sufficient support for the cause of action. Consequently, this legal stage is not a trial on the claims per se; the presiding judge must assess whether the case meets the criteria for class action certification. Judge G. A. Smith subscribed to the widely applied "plain and obvious" approach to this assessment, whereby claims are to be assessed generously, erring on the side of protecting the right to legal access, with certification denied only if it is "plain and obvious" that the claims could not succeed.

Class action certification for *Hoffman* was denied on May 11, 2005. In her 175-page decision, Judge Smith found no cause of action in the majority of the plaintiff's claims, relating to such common law torts as negligence, trespass, and nuisance. The judge did find that there might be some chance of a cause of action related to proper conduct of the defendants according to two environmental statutes (the 2002 version of the Saskatchewan Environmental Management and Protection Act and the Saskatchewan Environmental Assessment Act). Ultimately, however, the issue was moot as certification was denied on the basis that the plaintiffs had failed to prove their claim related to an identifiable class. The details of the judge's findings are described below.

With respect to negligence, the plaintiffs claimed that the defendants knew, or ought to have known, that the introduction of GM canola would infiltrate the crops of organic canola growers, and that they had a duty of care to prevent such an occurrence. Consequently, the defendants ought to have warned growers who purchased their GM products about the possibility of damage to neighboring crops and advised them of practices—such as using buffer strips, tarping trucks, and carefully cleaning machinery—that would reduce the spread of GM material. A significant issue for this claim was the timing of organic certification agencies' amendments to their standards to prohibit GM material. Further, the judge had concerns that the claim lacked evidence that the loss and damage to the plaintiffs' canola crops was foreseeable.

The type of technical logic necessary for legal claims included a detailed analysis of the applicability of previous case law to the claim at hand. For example, the plaintiffs claimed on the basis of *Rylands v. Fletcher* that the introduction of GM canola was a non-natural use of the land, with the defendants bringing onto their land something likely to do mischief if it escaped. This form of argument logically redirected the organic farmers' claim against their neighbors, however, instead of their intended target, the technology developers (something they were unwilling to pursue). Consequently, an underlying issue with the claims appeared to be the difficulty of fitting the novel issue of GMOs under the traditional actions available. In a similar vein, the judge expressed concern that the "relationship between the plaintiffs and the defendants . . . [failed to] . . . support a finding of relational proximity."[76] That is, the plaintiffs had to have had a sufficiently close and direct relationship with the defendants that they would have been mindful of their interests. Instead, the judge found the plaintiffs had "not alleged any relationship at all."[77] This gap between the creation and sale of the GMOs and their dissemination into the environment was a recurring theme, and a significant sticking point for the applicability of case law that evolved from more direct relationships, such as one person's oil tank leaking oil into another's well.

The nuisance claims rested on the premise that the GMOs had interfered with the organic farmers' use and enjoyment of their land. The counterargument claimed that the high standards of third parties (the organic certification agencies) caused this loss, not the technology producers. While questioning whether the loss had its roots in "hypersensitivity," the judge conceded that the novelty of the farmers' claim created difficult hurdles to overcome. Thus while it was not plain and obvious to her that the "use and enjoyment" argument lacked merit, she nonetheless found it failed because the defendants were not the cause of—and thus were not liable for—the nuisance. The defendants only sold the canola, which required the active intervention of farmers in order to create the adventitious presence; finding them liable would be the same as "holding the manufacturers of pesticides responsible for the nuisance caused by the harmful drift of the pesticide."[78] The sentiment that this would be a highly negative evolution of case law was strong: "[t]he implications of holding a manufacturer, or even inventor, liable in nuisance for damage caused by the use of its products or invention by another would be very sweeping indeed."[79] This interpretation overlooks the difference between, for example, finding a crowbar manufacturer liable for a murder, when there could be a non-murdering use for a crowbar, and finding the makers of GMOs liable for their dissemination (when they were used as intended), however. While a careful path must be tread in the application of case law, the above logic seems to give insufficient weight to the novelties the new technologies introduce.

Further to the novelty problem for the plaintiffs was the judge's apparent reluctance to contradict government approval of the technology. Judge Smith stated that there were policy considerations to her decision, given that the plaintiffs had federal approval to release the GM canola: "[T]he imposition by the courts of a duty of care not to release these substances into the environment would therefore appear to be in conflict with express governmental policy."[80] This reluctance to find liability for an approved product seems out of sync with the raft of other products that have run the gamut of liability litigation, with tobacco being an obvious case in point. It also appears to add an insurmountable level of difficulty to claims of this nature. Despite this reluctance to override government approval, Judge Smith did find some room for questioning the technology—and potentially lodging a claim against it—within two existing governmental approval processes.

Under the Saskatchewan Environmental Assessment Act, if the release of GM canola is construed as a "development" according to the meaning of the act, then the defendants would be required to conduct and submit an environmental impact assessment and obtain ministerial approval, or face liability for any loss, damage, or injury as a result of their development. The plaintiffs would first have to demonstrate that the release of GM canola is indeed a "development," however, the determination of which is very difficult and vague, even for key agencies, but it could include the following: unique features, resource use, emissions, widespread public concern, new technology, and a significant impact on the environment. While noting the difficulty of proving GM canola is indeed a development, the judge decided that it was "nonetheless impossible . . . to conclude that it is 'plain and obvious' that this statute does not apply in these circumstances."[81]

In 2002, Saskatchewan's Environmental Management and Protection Act [EMPA] was repealed and replaced with a new version. The organic farmers had made their claim under both versions, asserting that (depending on the version) GMOs are a pollutant/substance that has caused loss or damage, and that the defendants own/are responsible for the discharge and are therefore liable. The earlier version of the EMPA required that a "pollutant" be "discharged" into the "environment," with each of these terms having a very strict definition. Pre-2002, for example, "environment" included only soil, water, or atmosphere, and therefore precluded the genetic alteration of plants: post-2002, however, it included "organic and inorganic matter and living organisms," and even the "interacting natural systems and ecological and climatic interrelationships" that include the other components.[82] Similarly, the very broadly defined "substance" of the later EMPA is much more encompassing than the very restrictively defined "pollution" of the earlier version. Given this broadening of definitions, Judge Smith found that it was not plain and obvious that a claim under EMPA 2002 could not succeed. Unfortunately for the plaintiffs, this version of the EMPA came into effect after the alleged loss of organic canola markets. Consequently it was not applicable to that loss, but only to any post-2002 loss, such as cleanup costs for GM canola invading organic farmers' fields.[83]

While the judge found no cause of action under the earlier and stricter EMPA, there remained the potential for amendments to the claim that might support such a cause. The plaintiffs faced another hurdle, however. The basis of the EMPA argument rested on the contention that the defendants "owned" the GM material. This claim was most specifically aimed at Monsanto, whose technology continues to be controlled post-sale through their TUA. Surprisingly, however, Judge Smith found it to be without merit. Her perspective on ownership was particularly significant in light of the *Schmeiser* proceedings, and of the question of expropriationism:

> [It] is not reasonably arguable that ownership of a patent in the modified gene and enforcement of patent rights through "technology user agreements" are sufficient to constitute "ownership" and "control" of the "pollutant" (GM canola seed and resulting pollen) after the seed is sold to farmers and cultivated by them, as these words are used in the Act. The "control" asserted by the technology user agreement is not control of when and how GM canola is cultivated or harvested, but only control, or restriction, of the right to save and use seed from the GM crop.[84]

Thus, according to Judge Smith, the plaintiffs "do not reasonably support the conclusion that the defendants owned or controlled the 'pollutants' at the time they were discharged into the environment."[85] Given all the restrictions and the infringement liability over any unauthorized use of GM technology, however, farmers would likely be very surprised to hear that the defendants did not "own or control" it.

In short, the judge's decisions revealed a number of significant disadvantages facing the organic farmers' attempt to associate liability with the ownership that Monsanto had affirmed in *Schmeiser*: the difficulty of applying existing case law to a novel technology; the difficulty of attributing liability to technology developers when it is farmers who are in direct physical contact with the technology; and the contradictory perspectives on the meaning of ownership and control. These hurdles leave organic farmers with little recourse against the threat to their industry, outside of negotiating with their markets and certification agencies to reduce or eliminate their GMO prohibitions. Were the approach in the class action decision to be upheld in a trial, the combined impact of *Schmeiser* and *Hoffman* would be a perfect expropriationist pair: biotechnology developers are awarded the full benefits of ownership, but this award is not associated with any of the liabilities traditionally associated with ownership. At this level of proceedings, however, the merits of the case were only assessed to the extent of establishing the plausibility of a cause of action. Ultimately, the more formalistic aspects of the application—related to the identifiability of the class, commonality of issues, representativeness, and preferability of the procedure—are what prevented these issues from gaining full trial.

The most significant of these issues related to the difficulty of determining an identifiable class with common issues. There is sufficient variation in farming that it precludes easy identification of who makes up the class in the same way that shareholders in a stock market scam make an obvious class. The identifiable class in *Hoffman* would include those who lost their market and/or who suffered a loss due to cleanup costs and other restrictions. Given the removal of wheat from the case, the number of organic farmers who were directly impacted in this way decreased drastically. The documented number of organic canola growers was very low, although organic farmers contend that the industry was just taking off when GM contamination stopped it. A variety of factors affect what a farmer chooses to grow in any year, however, and many factors other than the market risks of GMOs could have motivated a farmer's choice not to grow canola. Further, attempts by the Organic Agricultural Protection Fund [OAPF] to solicit testimonies on the issue yielded little. This failure to identify a clear class was decisive for the judge's denial of certification:

Members of the class sought to be certified farmed at various times, in various areas of the province, in various circumstances, were certified by various certifiers with varying standards (both among themselves and over time) and sold or tried to sell produce into various markets with varying standards (both among themselves and over time). The proceedings in this case would, in my view, inevitably break down into individual proceedings, requiring full discovery rights and a trial of the factual issues.[86]

Given the importance of not allowing class actions to make defendants liable for those they have not impacted, the proper determination of the class is crucial. Certainly, there were reasonable concerns regarding the pervasiveness of the lost canola market. While the decision might appear reasonable for the specific case of organic canola, however, the underlying logic behind it would seem to preclude any future class actions by farmers, however central the crop in dispute. Farmers will always be situated on different land that is suitable for different crops in different weather, and will base their cropping decisions on a wide variety of factors, including their predictions about crop prices and their willingness to undertake the extra requirements of specialty production in return for higher prices, among others. These variables reflect the nature of farming, and yet they have been positioned as an insurmountable barrier to any class action against GMOs being initiated by farmers. The representativeness of the plaintiffs was another significant issue with respect to the potential for group action, as will be discussed presently.

Soon after their certification application was denied, the organic farmers applied for leave to appeal, which they were granted on August 29, 2005. Two groups also applied for intervener status on the case: Friends of the Earth and the Saskatchewan Environmental Society. On November 23, 2006, the appeals court judge denied intervener status to the applicants.[87] Class action certification was subsequently denied again in 2007, leaving the plaintiffs the option of pursuing their case through individual claims. In April 2008, the SOD announced that Hoffman and Beaudoin had elected not to proceed on an individual basis, and the first legal attempt to address the liability issue in North America reached its conclusion. For organic farmers, the case was about defending their industry against the ultimate act of expropriationism; for biotechnology developers, it simply provided organic farmers "a platform for their anti-biotech position" (Trish Jordan, spokesperson for Monsanto, 2005).

The Best Defense Is a Good Offense

Saskatchewan's two legal cases garnered a vast amount of media publicity and stirred up a great degree of controversy over the introduction of GM crops. Outside of these two cases—where the litigants are on the offensive in the court of public opinion—it is difficult to gain qualitative or quantitative data on Canadian farmers who have had any such contact or involvement with biotechnology developers. While those who accept settlement offers to avoid litigation are prevented from speaking out, the fact that Schmeiser's is the first court case suggests that the number of such Canadian interactions is considerably smaller than in the United States. These interactions do exist, however, and carry a significant level of threat. For example, one individual who received communication from Monsanto (but was not yet legally precluded from talking) declined even an anonymous interview on the basis that he preferred to "let sleeping dogs lie" rather than risk provoking Monsanto's wrath (telephone communication).

Some information does nonetheless reach the public, however. Schmeiser, for example, in his anti-Monsanto campaign, posts relevant material he receives from other farmers on his website, including three of what he terms Monsanto's "extortion letters." While the names have been removed from two of them, there is little doubt that Monsanto could identify whose letters they were. Thus publicizing the letters is an act of resistance in itself. In one, dated November 12, 1998, Monsanto informs the farmer that it has investigated and determined it has "very good evidence" to believe the farmer has 250 acres of RR canola.[88] The letter states that the company is willing to "refrain from commencing any legal proceedings" on condition the farmer (1) make a payment of $115/acre for the 250 acres (totaling $28,750), (2) commit to Monsanto's right to future testing, and (3) not "disclose the specific terms and conditions of [the] Settlement Agreement to any third party."

Another letter, dated December 17, 2001, states that Monsanto has determined that the farmer improperly planted 3,420 acres of its RR canola in 2000. It references the federal court case against Schmeiser to indicate that the precedent has been set for Monsanto to be awarded profits plus costs, and that while "[t]he amount of those costs have not yet been determined . . . Monsanto has reason to believe that they will be substantial." Given that the trial court assessed Schmeiser's profits at $19,832 for 1,030 acres of canola ($19.83 per acre), costs were an important economic threat. Monsanto offered to settle the disagreement "amicably" for an amount of $50/acre, totaling $171,000. A last letter, dated December 7, 2004, claims the farmer has 700 acres of unlicensed canola. While it references previous telephone communication, it is interesting that the letter itself—dated after the Supreme Court decision which awarded Monsanto no remedies at all—provides no concrete offer, but only suggests that a settlement plan "that will enable you to continue farming in the future" could be worked out instead of proceeding to court.

The farmers who testified in the Schmeiser trial about their experience with unwanted volunteer GM canola provide further insight into Monsanto-farmer interactions. Mr. Gerwing, for example, first made contact with Monsanto over his GM canola in 1998, when he saw a Robinson Investigation truck parked at the elevator:

> He was in there and I told him I wanted a leather jacket. I heard they give away leather jackets if they report that somebody is growing Roundup Ready without their—what you call it—fee for growing it.[89]

It is unclear whether this action was sardonic or indicative of the lack of awareness of the potential seriousness of the issue at that time. In any case, while testifying that Monsanto ultimately responded to the GM volunteers, Mr. Gerwing indicated some resentment over the company's attitude:

> But it's something I can't control. But when I phoned them to, what they were going to do about their canola, like I asked them, is it their canola or is it—like this fellow informed me that it was Roundup Ready canola after that and I asked him, "Well, is it my canola now or is it their canola? Or who owns the seed to it?" They said they own the seed, but they wanted me to do the spraying to kill it.[90]

Mr. Boser similarly testified that when the Monsanto representative confirmed the likelihood of RR canola, he became pretty upset:

> Well, basically, at that point I was a little upset to have his contamination on my land because I've never used Roundup Ready canola. And so I just said to him that if it's a Roundup Ready canola and it's—you claim to have the patent to the seed, I suggest you remove it from my property.[91]

Despite the fact that in both cases Monsanto ultimately responded, neither indicated the level of satisfaction over the issue that Monsanto's lawyers strove to establish. Mr. Boser, for example, testified that he told Aaron Mitchell from Monsanto that there were still pockets of GM plants growing that had been too small to pick. He indicated that Mitchell suggested 2-4-D wouldn't help, and probably cultivation would provide his best result.

> *Monsanto's counsel*: So Monsanto continues to assist you to deal with this problem to the extent it exists today?
> *Mr. Boser*: No, I wouldn't say that.[92]

Mr. Gerwin similarly indicated limitations to Monsanto's assistance with the management of their patented technology:

> *Monsanto's counsel*: And Monsanto representatives offered to follow up with you again this year to ensure there's no problem, correct?
> *Mr. Gerwing*: No, I haven't heard nothing from them.[93]

While insights provided by the settlement letters and the above testimony are rare, they do provide some indications of the nature of the interaction between farmers and Monsanto when there are issues related to their technology. Whether claiming patent infringement, responding to complaints about their technology, or rebutting liability for it, as in *Hoffman*, Monsanto appears to be in control at all times.

An Aggressive Defense

For Schmeiser, his case was very much about social justice, and he invested considerable energy publicizing what he felt were Monsanto's bullying tactics. He states that he felt his relationship with the company was much more than a legal battle, and that they were using all sorts of tactics to discredit him—for example, framing its interaction with him in terms of preventing his having an unfair advantage over those who paid the technology fee—if not financially break him. Schmeiser felt maligned, given his perspective that he was only practicing his normal farming practices:

> That was an effective way to get farmers to say, "Well, why should he get away with it." They are 200 miles away, what do they know about it. Monsanto has reps in every region. I don't have any. (SK#9)

Schmeiser had little doubt that he was going to use everything he had to fight back, and to try to preserve farmers' rights to save their own seed.

There is little ambiguity in the consequent crusade that Schmeiser launched against Monsanto during his trial. It is also clear, however, that his resistance did not abate at its close. Drawing on Monsanto's trial evidence that they were available to remove the unwanted presence of their patented material, Percy's wife, Louise, filed a small claims court suit against Monsanto in Humbolt, Saskatchewan, over the $140 cost of removing unwanted RR canola from her organic vegetable garden. The case was heard March 21, 2005. The suit again garnered a significant amount of media publicity (see, e.g., Hansen, 2004) and also renewed interest in the issue of power distribution between farmers and biotechnology companies: given that Monsanto had sued Schmeiser over the presence of RR canola in his fields, could his wife sue them for the cost of having to remove "their" canola when it was unwanted? Despite the publicity, however, less than a handful of people attended the actual hearing (author's notes).

The claim stated that volunteer canola had appeared in the Schmeisers' shelterbelt and in Louise's garden in the summer of 2002. Two applications of Roundup in the shelterbelt revealed it to be Roundup resistant. Louise Schmeiser sent repeat requests, by mail, to Monsanto requesting they come and remove the plants. When these requests were ignored, she ultimately had the plants removed and sent Monsanto the bill. The case was dismissed for a number of reasons. For one, Louise Schmeiser did not attend the trial as she was unwell and unable to testify on her own behalf. It is also no small point that the Schmeisers were actively engaged in litigation with the defendants at the time of the alleged contamination. Given its trail-blazing nature, and notwithstanding the low level of court, the case is important for continuing to tease out the evolving legal framework for the technology, specifically here with respect to the standards for documenting the presence of volunteer RR canola, and for Monsanto's obligation over these volunteers.

Percy Schmeiser thoroughly documented his interaction with the unwanted canola: he provided dates and photos of the canola in the shelterbelt and in the garden; he documented himself and another party spraying Roundup in the shelterbelt, and he provided follow-up photos post-spraying. He did not provide evidence on the concentration or rate of application of the Roundup, however. In Monsanto's defense, Mr. Ripley, a technology development representative for Monsanto, testified that the rate of application of Roundup was crucial as certain environmental factors can influence its effectiveness—consequently, a lab test was required to positively identify RR canola: "It was his opinion that spraying was not an accurate test to determine if Monsanto's gene was present."[94] The case was ultimately dismissed on the basis that it was inconclusive whether the canola plants in question were Roundup resistant, despite the fact that the Supreme Court of Canada in Schmeiser had found that "a canola plant that survives being sprayed with Roundup is Roundup Ready Canola."[95] The small claims court judge required a higher standard of evidence for the presence of RR canola on the basis of its existence being the primary issue of the case, and because Schmeiser had not denied its presence in the earlier Supreme Court case. It found:

> [I]n view of the absence of evidence with respect to rates of concentration and the environmental conditions during the spraying as well as the Defendant's evidence that only a lab test would be the ultimate determinant of whether or not the Roundup resistant genes were present in the plants I cannot conclude that the Plaintiff has proven that these are Roundup Ready plants on a balance of probabilities.[96]

By these standards, only a lab test, which farmers are unlikely to cheaply access, is sufficient for those seeking redress over the presence of RR volunteers.

A second significant issue was whether Monsanto had a duty of care with respect to removing any unwanted canola. As in *Hoffman*, the defense placed a heavy emphasis on the fact that Monsanto had been granted federal regulatory approval for "unconfined release" of its product. Consequently, the court could find no evidence "that there was, in fact, a duty of care to the Plaintiff to ensure that there is no unwanted spread of Roundup Ready canola plants."[97] While Monsanto had provided evidence in *Schmeiser* that it had responded to farmers' complaints of unwanted GM technology, this was taken by the small claims court only to indicate that Monsanto "voluntarily assumed" this responsibility.[98] If this court's perspective holds, not only will farmers have a very difficult time proving the presence of volunteer RR canola, but Monsanto will be under no obligation to remove it when they do.

This would seem to conclude Schmeiser's potential interactions with Monsanto, but it did not. Despite having quit growing any canola, Schmeiser alleges that in September 2005, his chemfallowed field was invaded by

RR canola volunteers. Monsanto agreed to remove the plants, but only on condition that Schmeiser sign their legal release form for the removal of unexpected volunteers. He refused, subsequently putting the document up on his website. It states in part:

> It is understood and agreed that the said delivery of product is not deemed to be an admission of liability. . . . It is further understood and agreed that the terms of the Final Release & Settlement of Claim shall be treated as confidential and shall not be disclosed to others without the written consent of Monsanto Canada Inc. and it is hereby declared that the terms of this settlement are fairly understood, that the amount stated herein is the sole consideration for this release and that the said product is accepted voluntarily for the purpose of making a full and final compromise, adjustment and settlement of all claims for losses and damages resulting, or to result from any of the matters referred to in this release. ("Release," Monsanto vs. Schmeiser, website)

In an October 2005 media article, Schmeiser estimated his cleanup costs to exceed $50,000, for which he intends to send Monsanto an invoice, and potentially file another lawsuit (Pratt, 2005). Trish Jordan, Monsanto Canada's communications officer, responded that "in this situation it would appear that Mr. Schmeiser is not really interested in assistance. He's interested in continuing his media campaign" (cited in Pratt, 2005, n.p.). This is very likely the case, but it does not render any more benign the inequitable power dynamics that Schmeiser helped reveal.

What has been the impact of Schmeiser's many years of resistance against Monsanto? Despite three levels of court, Monsanto's patent was upheld. There is no doubt that Monsanto will increase its enforcement efforts now that it has had its patent affirmed. According to Trish Jordan: "We did have a number of people waiting in the queue, but (Schmeiser) was the first case where we attempted to find out if the patent was valid" (cited in Lyons, 2001, n.p.). It is very likely that *Schmeiser* drastically reduced the damages Monsanto had hoped to receive in such cases. Depending on the size of acreage, of course, the greatest expense a farmer might face in litigation could be legal costs. This reduces the economic threat the company can wield in settlement offers. However, even if each side has to pay its own costs, as in the final *Schmeiser* decision, and even if future cases are unlikely to proceed as far through the courts, these costs can still be formidable. As Schmeiser noted:

Where it hurts an individual like ourselves in the decision is when the court ruled that we each had to pay our costs up to that point. Our costs probably was in the neighborhood of well in the excess of $400,000. (SK#9)

Schmeiser's legal actions also provided far greater exposure to numerous issues around agricultural biotechnology, and ultimately gave some clarity to what might constitute "innocence" for the courts. It is also likely that his campaign increased the exposure for related issues like the Seed Sector Review. For groups such as the National Farmers Union, who are active on the issue, the changes attempted by the review and by the patenting of GM seeds are part and parcel of the same drive to alienate farmers from their right to save seed.

The Organic Play

For organic farmers, GM technology went from an issue that was irrelevant to their farming practices to a pressing concern in just a few short years. Once faced with the prospect that this new technology was going to have a negative impact on their burgeoning industry, they fought for a way to influence the outcome. Opportunities were few, however, and ultimately the Saskatchewan Organic Directorate began to feel that legal action was the only route:

This was an issue back in probably '99, '98; we went to the provincial government lobbying, and said, "We're looking at if you guys don't do something about this GMO stuff"—we had a meeting in Regina and we were there, expressing our concerns, and basically threatening some kind of legal action. Of course the government never thought we would do it. (SK#6, Taylor, SOD president)

Members of the SOD were sufficiently concerned about the impact of GMOs on their industry that they felt they had to act, however. In this context, it needs little emphasis that *Hoffman* is not simply the result of two organic farmers who sought refuge in a representative organization. Instead, *Hoffman* resulted from a significant degree of deliberate action from an umbrella organization that wished to represent the concerns of its members more broadly and to avert further negative impacts from impending GMO introductions. Ironically, this broad-based concern ultimately hampered its acceptance as a class action.

In an effort to mitigate the impact of the litigation on the representative plaintiffs, and as advised by their counsel, the SOD and the representative plaintiffs entered into two written agreements: one was an "agreement to act as a representative plaintiff," which outlines the role of the plaintiff, asserts his agreement to attend court as required, and gives authorization for the SOD-OAPF to retain and instruct counsel on his behalf; the second was a "legal costs indemnity agreement," which protects the representative plaintiff from any award of costs which may be levied against him in the course of the litigation. The judge expressed concern that as a result of these agreements the plaintiffs "relinquished control" over the lawsuit. For example, while Hoffman was one of the twelve SOD-OAPF committee members, he had no independent right to conduct action outside of his role on that committee, and Beaudoin had never attended a committee meeting and did not even know the identities of all the committee members. According to Judge Smith, however, a representative plaintiff has the responsibility of prosecuting the lawsuit, and these duties cannot "be delegated to another party who is not answerable to the Court."[99] Consequently, she concluded that neither plaintiff could "fairly and adequately represent the interests of the class."[100]

In their defense, the SOD characterizes the relationship between the SOD and the plaintiffs thus:

> SOD is not suing Monsanto; the plaintiffs are suing Monsanto. We are facilitating the lawsuit. We have an agreement with the plaintiffs. . . . Both plaintiffs are on the committee as well. It is kind of a chicken and egg thing. There was a move afoot to take some sort of legal action, and we were the right responsible body to help make it happen. (SK#6, Taylor)

While it is understandable that the court might read the role of the plaintiffs as being a means to an end for the goals of the SOD-OAPF, it is nonetheless difficult to construe how this is negative to the interests of the class, or somehow less in their interest than such actions initiated by lawyers. The SOD believed that its members had a legitimate class interest and sought an available means of representing it through the legal system. The plaintiffs were amenable to tearing up the disputed agreements, but this was not sufficient to change the perspective of the court.

The benefits of a committee-run class action lawsuit are large for the plaintiffs, however, considering the toll such litigation would otherwise take on an individual plaintiff. If the only way such litigation could proceed was through an individual plaintiff, it would have a very small chance of occurring—indeed, the legal chronology ultimately ended on this basis. Considering organic farmers' limited options for opposing GMOs or otherwise assigning liability for them, litigation appears to offer a very rare avenue for action. It is also an avenue that is unlikely to proceed without class action certification:

> Not being certified would probably spell the end of the litigation in all likelihood. The individuals do not have the wherewithal to pursue it on their own, and they are not likely to get the support they need just to pursue it through actions. It just becomes non-feasible to do it. There is a major concern about cost. (SK#1, Zakreski, lawyer for Schmeiser and Hoffman)

Further, the case was about much more than canola, and the SOD's determination to fight was directly related to its concerns about future GM introductions. While wheat was withdrawn, GM alfalfa hovered on the horizon. As long as litigation remained a possibility, the SOD was committed to prevent GM contamination of their industry. For example, with respect to GM alfalfa, the president of the SOD stated:

> We have written to Monsanto alerting them that we are aware of their possible intention to introduce GMO alfalfa, and we will probably— if they try it—we will probably try to take legal action against them through the injunction process. (SK#6, Taylor)

Ultimately, with the failure of their class action, the SOD turned to other strategies, and in April 2009 it announced a widespread appeal for more groups to join the over 80 groups already signed on to their "No to GMO Alfalfa Campaign," created in alliance with the Canadian Biotechnology Action Network [CBAN] (CBAN, 2011).

While establishing liability would have been a monumental victory for the SOD, achieving the lesser goal of requiring compliance with any of the environmental acts would have provided a significant avenue for the SOD's concerns to be heard in the future.

> If we can attach any area of liability, under the environmental acts, I mean, I'm hoping that this is what will happen. That from now on if we get something then they have to do an environmental assessment before they release it. That for me is the ideal situation, even if we don't win

damages or this and that; I mean we're not in it to make money. . . . if we can get it so that they don't just release GMO alfalfa, or before they have the proper hearings, just like a hog barn or anything else. (SK#6, Taylor)

Such an outcome could have forced the release of any new GMOs to be subject to a number of considerations and public hearings, much like those desired by the opponents of RR wheat. In point of fact, it is just this kind of procedural hurdle which subsequently resulted in revocation of the approval for RR alfalfa in the United States in 2007, when a federal district court judge ruled that the USDA needed to complete an Environmental Impact Statement before any further RR alfalfa could be planted. The ruling was upheld on appeal in 2009. Perhaps due to the RR wheat lesson, commercialization in Canada is unlikely without approval in the United States (CBAN, 2011).

While the SOD lost its bid for class action, the impact of the SOD's attempt, like that of *Schmeiser*, reverberated beyond the courtroom. For one, it raised the issue of liability into the light, and rendered visible an otherwise invisible externality of GM technologies, providing fuel to the various anti-GM campaigns of environmental and other NGOs. For another, the case highlighted aspects of the technology that require an unavoidable social decision about property rights that cannot be resolved by preexisting legislation (in the manner that "substantial equivalence" regulates health and safety decisions under preexisting statutes). Once raised by *Hoffman*, the issue of liability at least earned an uncomfortable mention at the agricultural policy table, if not an actual seat. As the plaintiffs' solicitor put it prior to the case's conclusion:

If the case is successful, it's significant because it established liability, and that would be of interest to legislators. But if there is no liability, then you at least know that the existing law can't address liability problems with these crops, and the government should be looking at enacting legislation to deal with it. (SK#1, Zakreski)

Of course, legislators have been known to ignore such regulatory needs, particularly on such "hot" issues as GM technologies, where any sort of action will likely trigger negative attention.

Conclusion: Hoffman v. Monsanto v. Schmeiser

Canada is clearly moving from its early reluctance to patent life forms to an approach on intellectual property more in keeping with that of the United States. Changes to the laws around patentability evident in the shift from *Harvard College* to *Schmeiser* are consistent with other drivers in agriculture—such as the Seed Sector Review and the growing incidence of seed contracts—demonstrating a persistent shift toward fully commodifying the seed. The litigation in Saskatchewan suggests an impressive legal fiat of expropriationism. In this assessment, *Monsanto v. Schmeiser* and *Hoffman v. Monsanto* cannot be considered separately; rather, the two represent an integrated package of property rights decisions. Contrasting the perspective on ownership in *Schmeiser* with that in *Hoffman* and even in Louise Schmeiser's small claims court decision, it appears that corporate ownership over patented GM seed is seamlessly maintained through its regeneration and wherever it wanders, but stops at the first point of sale with respect to liability. If this were to be upheld, it marks a flawless transition to expropriationist accumulation: biotechnology developers hold the benefits of ownership without many of the normal responsibilities, while farmers bear the costs. In Hoffman v. Monsanto v. Schmeiser, there is little ambiguity as to who has come out the winner.

Schmeiser's legal battle with Monsanto did much to expose the expropriation of farmers' property rights associated with patented GM crops, although the outcome of the case was certainly affected by Schmeiser's awareness of the patented material. Thus a true case of an "innocent" infringer remains to go before the courts, where "innocence" connotes lack of knowledge and/or a rebuttal of the presumption of use raised by possession. Many questions remain with respect to this scenario: What percentage of GMOs constitutes "occasional" presence in a farmer's field versus an infringing presence? What are the grounds for countering the presumption of use? Should farmers spray to assess for Roundup tolerance, or not spray to avoid determinations of use? If farmers won't sign a release to get Monsanto to remove unwanted GM technology, will they be vulnerable to infringement suits? Even if such uncertainties are ultimately resolved in a farmer's favor, the expense of litigation means a negative impact for any farmer facing these issues in court, regardless of legal outcome.

Certainly, technology developers are unlikely to wantonly pursue their customers. There is no doubt that abuse can and does occur, however, and given the imbalance of power, legal uncertainties are troubling. Further, the lack of independent sampling and handling procedures not only leaves it in the hands of technology developers to bring the accusation of infringement, but also to gather, test, and hold all the evidence. Farmers could commission independent tests, but this is costly, and its importance is most evident only once it is too late. Given that the benefits of the patent accrue to its holder, there are also equity concerns in a construction of infringement that puts the onus on farmers to be the guardians of wayward GM technology, particularly when *Schmeiser* is set in contrast to *Hoffman*. The scene set by the two cases certainly suggests that developers get to have their cake and eat it too with respect to ownership and control.

Last, it is important to note regarding the evolution of such case law in Canada that as Schmeiser was not a Monsanto customer, he did not sign a TUA, and its contractual provisions did not come under legal scrutiny. Such issues have advanced much further in the United States, where numerous questions have been raised regarding the contract's restrictive terms. When such a suit occurs in Canada, there will likely be an equivalent airing of important social issues (see, e.g., Kaiser, 2005).

For those who have attempted to resist the technology and its proprietary aspects—what I term expropriationism—the strategy has been twofold: one part litigation and one part capitalizing on the publicity raised by the litigation. While the rapid expansion of biotechnology in Canada has been an economic agenda, with relatively little attention given to regulating risks or addressing the new proprietary issues, NGOs excluded from policy processes have capitalized on these court cases to manage a sustained public pressure on the industry. Although some might disagree with the extent to which Schmeiser took his anti-GM crusade, it is also certain that without Schmeiser, many changes led by GM technology, including some questionable power aspects, would have occurred beneath the radar. Similarly, *Hoffman* raised the ire of those who thought the loss of a tiny organic canola market did not warrant such a legal show; in consequence of it, however, ownership of GM material is irrevocably associated with the question of liability. In short, these lawsuits increased the transparency of the changes that are occurring and opened up the potential for publicly motivated policy development.

CHAPTER 6

From When Cotton Was King to King Monsanto

A lot of folks in Mississippi, two generations ago, were living on very hard scrabble farms, doing everything by hand. I've got kinfolks who are long removed from farming, but they grew up in it, and couldn't wait to get out, and they still think only in terms of every new technology that has ever been available to agriculture—from tractors, herbicides, anything that comes down the pipe—its only possibility is in being a wonderful thing that will help you from having to work as hard. . . . Intense back-breaking labor is bad; therefore all technology is good.

MS#23, MEDIA/ORGANIC PRODUCER

That's the least of my worries, is agriculture biotechnology.

MS#30, MISSISSIPPI DEPARTMENT OF AGRICULTURE AND COMMERCE

I don't see it as any different of a tool of conventional crop breeding or inorganic fertilizers that came on in the probably early twentieth century. Or moving from a mule to a tractor was technology. Or moving to an airplane was technology improvements. I don't see that it's any different. You have to continue to evolve. That's just part of it.

MS#3, GM PRODUCER

Introduction

The United States is the leader in biotechnology with respect to its early experience, its GM-dedicated acres, and its extent of adoption in key agricultural crops such as soybeans and cotton; as a whole, it is indisputably pro-biotechnology. There is considerable differentiation within it, how-

ever, given its immense geographic and cultural domain. Anti-GM protests and local initiatives to ban GMOs—in Mendocino County, California, for example—are on one extreme. The state of Mississippi would
seem to occupy the opposite extreme.

Mississippi is in the U.S. Deep South, its southern end bordering the
Gulf of Mexico. The state has a population of 2.8 million spread over 30
million acres, much of which is forested. Average monthly temperatures
range from a low of 34.9 to a high of 92.5 degrees Fahrenheit (Netstate
.com, "Mississippi"). Summers are long and hot, and winters are short
and mild, allowing for an extended growing season. At one time, when
slaves were considered property and cotton was king, Mississippi was
counted among the richest U.S. states. Now it is the poorest of all U.S.
states, though much of the social, cultural, and economic infrastructure
of its cotton days is still evident.

Almost 40% of the state of Mississippi is farmland (USDA, Economic
Research Service [ERS], "State Fact Sheets"). In 2004, Mississippi had
42,200 farms (ibid.). The average farm size is 263 acres, although there is
great variation, ranging from the smallest low-resource farms to the massive cotton farms in the delta region. Over 50% of farms are under 100
acres, and 90% are under 500 acres (ibid.). Most of the very small farms
are subsistence production for the state's poorest. The bulk of Mississippi's agricultural production comes from a very small proportion of
farms, however: in 2002, 5% of farms were accountable for 75% of the
state's agricultural sales (USDA, National Agricultural Statistics Service
[NASS], "Mississippi State"). Geographically, row crop agriculture in
Mississippi is divided between the delta and the hills areas. Along the west
side of the state runs the Mississippi River, providing the rich soils of the
Mississippi delta. In the wide, flat delta region, soil is good, water is plentiful, farms are expansive, and cotton still dominates. This region is where
the largest farms of Mississippi are located, easily 5,000 to 10,000 acres.
To the east of the delta are the hills, only qualifying as such by contrast to
the flatness of the delta. Farms are smaller in the hills, more diversified,
and broken by forested areas. While not a focus here, inklings of attitude
differences toward biotechnology were evident between farmers in the
hills and the delta.

Post–World War II, agriculture in Mississippi became more diversified,
although both row crop agriculture and cotton have remained important
to the state economy. Currently, agriculture more broadly speaking—
including poultry, forestry, catfish, and cattle, as well as row crops—is the
number one industry in Mississippi, worth $6 billion in 2005 (Mississippi

Department of Agriculture and Commerce [MDAC], "Mississippi"). As such, agriculture provides direct and indirect employment to 30% of Mississippi's workforce and is significant to all of the state's 82 counties (ibid.). Cotton now ranks after poultry and forestry, but it is still the number one crop, bringing in $598 million in revenue each year (Mississippi State University Extension Service Website, "Crops: Cotton"). Leaving aside animal husbandry and forestry, the main agricultural products for Mississippi (in order of production value) are cotton, soybeans, rice, hay, and corn (USDA, NASS, "Mississippi State"). The extensive crop rotation evident in Saskatchewan is not as prevalent in Mississippi, although some certainly practice it. Cotton, in particular—especially on the larger farms—is typically grown year after year, with some acres sometimes put to soybeans and corn to hedge for market fluctuations. Rice is usually grown in a different area, where the soil is heavier clay, and is sometimes grown with soybeans.

Given the genetic traits that have been the focus of biotechnology development and the crops in which they have been introduced, Mississippi farmers have had a significant amount of opportunity to access the technologies. Of the state's top five agricultural crops, GM varieties are available in three: cotton, soybeans, and corn. While GM rice is not yet commercially available, Ventria's attempt to introduce pharmaceutical rice in neighboring Missouri made it an important topic in Mississippi (Elias, 2006). Cotton, soybeans, and corn are all available with Monsanto's Roundup Ready trait. Monsanto began the launch of its RR products with RR soybeans in 1996, followed by RR cotton in 1997, and RR corn in 1998 (Monsanto Company, 2005). In 2004, Bayer CropScience introduced an alternative, with the launch of Liberty Link cotton. Monsanto introduced insect protection for crops using Bt in cotton (Bollgard cotton) in 1996 and in corn (Yieldgard corn) in 1997 (ibid.). Stacked RR and Bt varieties became available in cotton in 1997 and in corn in 1998. While reportedly on the very near horizon, in 2005 there was still no competition with Monsanto's Bt system available in Mississippi. At that time, also, Bayer's alternative to Monsanto's RR cotton had only captured an estimated 3% of the market. Consequently, discussions of transgenics in Mississippi are essentially discussions of Monsanto's transgenics.

Once introduced, GM soybeans and cotton were adopted rapidly, though in the early years farmers reportedly balanced the convenience promised by the technology against hesitations over a perceived "yield drag" (decrease in yields), potentially caused by the technology's introduction into older crop varieties. Within a few years, GM technology was available in newer varieties, however, and debates about yield drag abated. By 2005, when the interviews were conducted, and barely 10 years since biotechnologies' introduction, 96% of the cotton and 96% of the soybeans grown in Mississippi were a transgenic variety of one type or another (USDA, NASS, 2005, "Acreage"). While corn statistics are not available, adoption of GM corn has been considerably slower than for cotton and soybeans. This is because the benefits of GM corn have been lower—due to the greater availability of alternative management techniques, higher yield drag, and less corn borer insect pressure in some regions of Mississippi; however, adoption of RR corn is reportedly strongly on the rise due to drift, as will be discussed.

This chapter draws on interviews with agricultural stakeholders and knowledgeable informants to provide insight into how biotechnologies' introduction has affected agricultural production in Mississippi. It focuses mainly on the responses of 33 of these interviews, which included 16 agricultural producers (3 organic, 10 GM, and 3 conventional); 12 with agricultural organizations, government representatives, and other stakeholders; and 5 with knowledgeable informants such as academics and media representatives. The remaining interviews, while contextually important, predominantly referred to the litigation that will be discussed in Chapter 7.

Evolving Technology in Mississippi

Because of the heat and humidity of the South, insect and weed management is a crucial part of Mississippi farming. Without the long, killing frost of winter, weeds and insects are a much greater pressure in Mississippi than they are in Saskatchewan. Any tool that helps with these pressures is going to be given strong consideration by the southern farmer. As the statistics indicate, adoption has gone much further than mere consideration; it is almost complete in cotton and soybeans. Technologically speaking, for the vast majority of Mississippi farmers and agricultural stakeholders, GM technologies were viewed with unmitigated approval. Those who used them were virtually unanimous in their praises of their physical attributes. One user, describing RR technology, captured the perspective of most of the producers I spoke to:

It's so easy. It's such a simple, easy thing to do; it's just wonderful. (MS#34, GM producer)

There are dissenting voices, of course, but these are extremely few and far between. What little dissention can be found primarily comes from those who operate outside of row crop production—notably small-market producers who are more closely tied to, if not actually involved in, sustainable agriculture movements. For those who love the technology, praise is extensive. Transgenic crops are reported to reduce farm management, labor, and energy requirements; decrease risk; and increase farmers' free time. Much less unanimously, some also argued for its economic benefits. While the vast majority of farmers and agricultural stakeholders unequivocally appreciate both the Bt and the RR technologies, these two technologies were adopted, and continue to operate, under different dynamics.

Roundup Ready technology is fundamentally a tool of efficiency and convenience. All the benefits of weed control evident in RR canola in Saskatchewan are maximized in the Deep South, where weeds and insects abound. Roundup Ready technology is used by 96% of soybean farmers and (between RR and RR/Bt stacked) 82% of cotton farmers in Mississippi (USDA, NASS, 2005). Extension agent estimates indicate that 50% of the corn in Mississippi is now Roundup Ready, after a recent acceleration in adoption in the last few years due to "drift" problems (airborne chemicals moving off target). Because the RR trait allows herbicide to be applied over-the-top of crops, it can be used in no-till farming, thus conserving moisture and drastically reducing the amount of cultivation a farmer has to do:

> I mean it used to be that we cultivated cotton. We had cultivators and ploughed the ground, and we'd make two or three or four trips with the cultivators. . . . But now we plant it and spray with it, boom. . . . and maybe come back and spray it one more time, and then maybe once more with a lay-down rig to put a chemical down, to keep the cotton once it gets big to keep the weeds from coming up all through. . . . It just makes it so much faster and so much easier to do. (MS#3, GM producer)

Ease of use is a management issue: as one farmer in Mississippi explained to me, when a farmer in the delta says he is going to *do* something to his crops, he means he is going to have it done. Reduced cultivation means farmers need less labor, which is not only expensive but is reportedly difficult to maintain in Mississippi, despite its status as the poorest state. Further, the nature of RR crops means that less care has to be taken to prevent damage to the crop, which up to the fourth leaf is not harmed by the herbicide, thus further reducing management concerns. This also means that if poor weather conditions prevent ground rigs from getting on the land to spray, the chemical can still be applied by air. Fuel and equipment are two additional high-cost inputs in agricultural production that are affected by RR crops: fewer trips to the field means less equipment is needed and fuel costs are drastically reduced — particularly significant in the context of high oil prices.

Despite these apparent cost reductions, the economic benefit of the RR technology is as ambiguous as in Saskatchewan. The economic benefit is taken as a given by some, who hypothesize that with fewer weeds they are getting more yield, which necessarily increases profits. Others have put pencil to paper and calculated no difference in profit given their high input costs. There are indications that the economic benefits accrue to those who capture the technologies' efficiency improvements by farming more acres. As explained by an agricultural economics expert, even with yield drag, which was an issue in the early years, the ability to increase acreage under the RR system maintains a producer's profit margin:

> If you could farm x acres before, you can now farm x plus 15% because the Roundup technology allows you to expand. And so, if you are making a dollar before per acre and now you are only 99%, but you've got 15% more acres, so that's what's led to a lot of the adoption of the Roundup in terms of soybeans producers. (MS#35, agricultural expert, economics)

Some evidence of this could also be found in conversation with producers:

> I wouldn't have taken [the extra fields] on, probably, if it wasn't for Roundup. I wouldn't have had the time or the extra money or the labor. (MS#6, GM producer)

Obviously, not all farmers can capture these benefits, and those who cannot capture them by increasing their farm size will be at a relative disadvantage, perhaps eventually being forced out. This impact of the technological treadmill on farm size is a familiar but important trend in the political economy of agriculture. The economics of biotechnologies will be discussed further presently. Irrespective of economic benefit, however, Mississippi producers had no doubts as to biotechnologies' overall merits: the contrast between farming before and after RR weighs in the background of conversations about the technology. One GM cotton farmer, for example, articulated the difference by referencing his fields, where weeds could be seen rising between the cotton plants. With RR crops, this only required an application of Roundup; in the past, it would have required a difficult balancing act, eradicating the weeds without damaging the crop.

> Twelve years ago, if this looked like this, I'd have been having a heart attack right now. Bad as it looked right here, I'd have been having a fit that we had to cultivate. Try to spray up under it an' everything. That's one thing, [the RR system] eases your mind quite a lot. (MS#4, GM producer)

Similarly, Monsanto's Bt technology is helpful for "easing the mind" for Mississippi farmers, but they have an even more powerful relationship with Bt. Between strictly Bt varieties and Bt/RR stacked varieties, the Bt technology is used by 73% of cotton farmers (USDA, NASS, 2005) and an estimated 30% of corn farmers (Bt is not used in soybean crops). The Bt toxin protects corn plants against pests such as the corn borer and cotton plants against the bollworm and budworm, the number one pest problem for cotton. In 1995, Mississippi cotton crops suffered a severe infestation of heliosis (the species of worms which includes bollworm and budworm). Farmers had to invest a huge amount of labor and chemicals in an effort to avert the damage, with no guarantee that they would not lose the crop anyway. Ultimately, the 1995 infestation was so devastating that many cotton farmers did not survive, and for those who did, cotton farming became a high-risk activity. It is in this context that insect-resistant Bt crops were introduced. The new technology had already been poised for launch prior to the outbreak. After it, farmers were reportedly so keen for the Bt technology that they pressed for its early release and planted all they could get their hands on.

Following are three typical experiences with the outbreak and Bt cotton:

> They knew we had a disaster in '95. It was a blessing when it came out. I mean, nobody was fired up about growing cotton in '96 after that big disaster we had out here. . . . It put a lot of people out of business in this area, it sure did. The Fourth [of] July we thought we had a good crop. We thought we had a pretty good crop. By the 10th of July we had that

next flight of moths come in and couldn't kill the worms. . . . you took to spraying and the next day the crops all covered you know it's over with. We picked anywhere from 300 to 400 pounds of lint. We going to pick 1,000 any other time. Sometimes more . . . I just don't think we'd be in the cotton business today if it wasn't for the Bt. (MS#4, GM producer)

Bt after '95, I don't know how much cotton would have been left in Mississippi if we had another year like that in '96. I mean, people were losing $200 to $250 per acre. It was devastating. When you have a loss like that, it takes 8 to 10 years to get over it. Bt really made a difference. Now, somebody in north Mississippi, or north Arkansas, or the Bootheel of Missouri might say RR may have a weed that fit it, and they might not have the trouble with worms that we did, so it might not have been the critical part. But right here I'd say it was the critical part of us staying in production. (MS#3, GM producer)

Prior to the Bt cotton we were having to spend up to $150 an acre controlling tobacco budworms. It was unacceptable. We couldn't continue to do that. People were going broke right and left. . . . When I say spending $150 an acre controlling tobacco bud worms, we were not controlling them, so we were having the damage as well as tremendous cost. So it came along at a time that it saved the industry, in my judgment. Now [GM technology] could get out of hand so it could kill the industry, if it wasn't contained, so to speak, cost and whatever. But there's no doubt that the technology . . . I mean, we could have lived without the Roundup technology, the glyphosate, but the cotton industry, I don't know if it could survive. (MS#34, GM producer)

Consequently, for many in Mississippi, Bt is not just another improvement to cotton farming, but the salvation of cotton farming. It provided the means to stay in business for those who survived long enough to gain that technological foothold against nature's destructive tendencies. By all accounts, this foothold has been extremely successful—so much so that heliosis is no longer the number one pest in cotton production:

When we know we have that protection on the front end, we know that we're not going to get significant damage at any time in the year from heliosis damage, which has been, up until the introduction of Bt, our number one pest in cotton. . . . [Bt] has actually changed the dynamics of our pest complex. Whereas heliosis used to be our number one pest concern, now it is what we call secondary pests, such as plant bugs. (MS#25, agricultural consultant, cotton)

Worm control is a variable cost, with some years facing high infestation and others low. As a pesticide that is incorporated in the plant, Bt offers full protection all the time, not just when the toxin is externally applied. It thus acts as an insurance package of sorts—for a price. It also reduces the amount of management required, for example, to regularly check fields for evidence of worms (which need to be sprayed before they burrow where sprays cannot reach). Further, the in-plant pesticide reduces chemical applications, thus again reducing costs related to trips to the field, fuel, labor, and equipment.

Few technologies will match the favourable conditions for introduction that Bt faced in Mississippi. Even without this industry-saving launch, however, the physical attributes of both RR and Bt technologies received only praise from interviewees. While drift is an increased management issue, as will be discussed, less management is required for most other production aspects. Overall, GM technologies are reported to take many of the risks out of farming—some of them significant, as in Bt—and to make production more manageable; essentially, they remove several important risk variables from the equation of successful farming.

A number of farmers also saw environmental benefits to the technology. While the RR system still requires herbicide application, farmers cite Roundup as one of the more benign herbicides. The main environmental gain, however, is seen to be the reduction in pesticides, the use of which is reported to have drastically declined since the introduction of Bt (particularly in contrast to 1995). Worker safety was another issue raised, as reduced chemical applications also reduced workers' chemical exposure. To be sure, contrasting examples of negative environmental impacts can be readily found on the Internet, as can convincing evidence of the temporary nature of any reductions in chemical use.[1] Nonetheless, the point here is that many of the farmers in Mississippi who use the technology see it as a benign alternative to conventional farming practices, in addition to providing them with vastly greater control over farm management. It would seem that in Mississippi, environmentalists' Frankentechnology is actually a princess in disguise.

The Drawbacks

Resistance and Drift

Princess or not, GM technologies are not completely problem free. Agricultural stakeholders raised two concerns related to the technologies' physical aspects—drift and resistance (by both insects and weeds)—the latter of which was really a concern about maintaining the continued viability of the technologies, rather than a concern about the technologies themselves. Strong negative sentiments come to light over the nonphysical aspects of the technologies—such as their price, the concentration of ownership, and farmers' current and projected relationship to GM developers.

With respect to insect resistance, Bt technology is sold with refuge requirements written into Monsanto's Technology Agreement [TA],[2] requiring a certain percentage of non-Bt crops to also be grown in order to prevent insect resistance from developing. These requirements differ depending on region—for example, refuge requirements for Bt corn are higher in cotton-growing areas, such as Mississippi (50%), and lower in corn-growing areas (20%)—and a farmer can select different strategies, varying refuge percentages with different cropping configurations. By most accounts, Mississippi farmers accept this interference in their production habits as a necessary constraint for maintaining the benefits of Bt technology. Refuge requirements do have some detractors, however; they argue, for example, that the requirements are an unnecessary burden given there are sufficient natural refuges in existence. None of the producers I spoke to made this claim, but it was raised by others, such as an expert in agricultural economics. Producers themselves appeared to respect the requirement and none reported any signs of insect resistance.

Weed resistance is another issue, however—one in which the experiences of producers are beginning to resonate with the warnings of environmentalists. Many farmers reported some experience or knowledge of developing resistance in weed species. Mare's tail, in particular, was a growing problem. While not outright resistant, according to some, it was increasingly difficult to control and required repeated chemical applications. For the most part, weed resistance was discussed as a significant future concern, but it was already raising immediate concerns for some:

Mare's tail, which was no problem whatsoever for years, all of a sudden it has become a major factor. . . . What a lot of people are doing right now who have mare's tail, they are having to do something that they never dreamed they would do again, they got people out there chopping the stuff down. . . . There's people around here who have major problems. And if they allow it to make seed this year, they might not be able to farm the ground next year. I mean it's serious. Whoever in the world thought mare's tail would be a problem, but it is. It really is. (MS#34, GM producer)

Interestingly, despite this resonance with environmentalists' perspectives on weed resistance, in the agricultural community it is often considered a normal part of production: resistance occurs and new chemicals are developed. Therefore, in the case of RR crops, for example, resistance is something to be delayed, if not actually avoided, in order to protect the functioning of the technology. Specific to GM technologies in Mississippi, however, is the concern that farmers are vulnerable in their dependence on a limited source—Monsanto—to provide the solution to future resistance problems.

Drift—or off-target chemical application—was another issue that arose out of the technology's physical properties. Drift has particular salience in the Mississippi delta, where wind can pick up and carry a chemical for a mile or two onto another's crops. Drift issues preceded Roundup, of course, and even now remain issues in non-GM crops. The widespread adoption of RR technology and the consequent widespread post-emergence application of Roundup and other glyphosates drastically exacerbated the issue, however, particularly when the technology was first adopted and awareness was far lower:

With Roundup, I think when it first came out, and this is my opinion . . . people weren't used to spraying Roundup as widespread as they were, and we may have had a little problem, but now it's cut down. (MS#30, Mississippi Department of Agriculture and Commerce)

The fact that RR crops are resistant to the herbicide meant that those growing them could relax their care around chemical applications somewhat, although this may not have been the case for their neighbors. Further, RR growers use Roundup at a time when conventional growers wouldn't, for fear of damaging their crops. This later application makes Roundup drift particularly dangerous to non-RR crops. While drift damage can be severe, it can also be quite subtle, resulting in a hard-to-verify loss in yields. Nonetheless, drift complaints became significant enough that the problem ultimately instigated new regulations designed to reduce drift: regulations controlled application timings and wind speeds; aerial applicators were required to log where, when, and at what wind speed they applied chemicals (to aid in assessing liability); and education programs were launched to help farmers reduce drift through drift control agents, proper spray tip sizes, and related means.

Overall, the new regulations appear to be working and drift problems have decreased. Given the nature of the technology, however, even with the new sensitivities, problems can arise:

> You can only spray [RR cotton] to a certain size, leafs, and it was so wet in June you know, a lot of cotton was almost at the 5th [leaf], and they couldn't spray it with the ground rigs, so they had to spray it by airplane, and it stirred up some controversy. . . . The farmers had their backs against the wall; their herbicide product was Roundup and all of a sudden guess what, they couldn't spray. Some people pushed the envelope on that one. They pushed it enough that I'd say maybe there was some damage. Maybe there was some drift damage. But the ag. pilots are the ones that have to—you know, it's a hard decision, he has to tell his customer, "No, no, no, no, no." (MS#33, seed dealer)

Drift regulations can obviously be quite burdensome to producers, and there are strong indications that they are taking measures to prevent the need for further government intervention. Farmers repeatedly expressed to me an awareness of their neighbors' farming practices, and that they communicate with them as a matter of course in order to avoid drift problems.

> Everybody pretty well wants to know what kind of crop their neighbors are doing. Everybody that I know is . . . with cotton I don't know anybody who plants conventional cotton anymore. . . . so you just have to ask are your beans and [is] your corn Roundup Ready? (MS#6, GM producer)

> We'll ask everyone around us: is this Roundup Ready or not Roundup Ready? We are being more cautious now with it. We aren't having the problems we were having two years ago. We were all just learning two years ago. We'll call each other up. If we have something that isn't

Roundup Ready, we'll be sure to tell the neighbors, watch out for this field. . . . We are getting the situation under control. It's not a problem now than when it first came out. (MS#24, GM producer)

Prevention of damage through good neighborly relations can only be a benefit to all. On the other side of the equation, however, are indications that some reductions in damage reports result from farmers declining to report damage. On more than one occasion, I was told that unless damage is extreme, farmers keep mum following a code of deference to the damage they may one day inflict on their neighbor—and in an effort to stem further application restrictions. A more worrisome trend, considering the concentration issues to be discussed, is farmers forced into GM adoption in order to prevent such damage. For example, the adoption of transgenic technology has not been as swift in corn, in part due to the availability of good conventional chemical strategies and in part due to a perceived problem with yield drag. Drift has been a particular issue for corn growers, however, both because the amount of conventional corn has remained high and because corn is highly sensitive to glyphosate. Consequently, preventative adoption of RR technology is reportedly now increasingly occurring:

A lot of growers have switched to utilizing RR corn to protect themselves from drift issues or drift problems, even though they had more hybrid choices available, and maybe a slight yield advantage with conventional hybrids. (MS#21, agricultural expert, corn)

A lot of people in the delta, I mean there's so many RR varieties now that a lot of people routinely choose RR varieties just to prevent getting drifted on. (MS#3, GM producer)

And there's one disadvantage now, so many people are using the Roundup, you almost have to make sure you use the Roundup as an insurance policy so your cotton won't get killed if you don't have it. Corn especially. We were forced into planting Roundup Ready corn, as I was getting so much crop damage. (MS#24, GM producer)

The problem of drift is far greater for those in the delta than in the hills area, where farms are smaller and more divided by natural barriers. It is also far greater for corn growers, although others also risk drift damage if they grow non-RR crops. The issue is significant enough that it potentially poses a barrier to competition in the Mississippi delta. For example, one farmer I spoke with who had land in both the delta and the hills area grew Liberty Link cotton in the hills, but only RR cotton in the delta because of drift. Monsanto's was the first technology to come out: now that it has been widely adopted, non-RR crops are a crop-loss risk for their growers and a liability risk for their neighbors. This is an issue for conventional crops, but it also inhibits the adoption of other, non-glyphosate-based, GM varieties. Where Liberty Link is planted alongside RR, for example, there is the potential for drift damage. The incentive for staying with the widely adopted Roundup is therefore significant. As one farmer explained it to me:

> If you plant Roundup Ready like everyone else has got, you aren't going to have any problems. (MS#4, GM producer)

Drift, then, is not simply a management concern, but contributes to monopoly control, an issue already of some salience to GM production in Mississippi.

Drift and resistance, however, are primarily considered things to be managed in order to maintain the technologies' effectiveness. It is only over the technologies' nonphysical attributes that stakeholders begin to express significant negative comments. In addition to the occasional reference to export restrictions or market concerns, negative comments inevitably turned to the issue of the technologies' cost, which was the number one concern. While cost was a significant, practical problem, it was intimately linked to the issue of control. Control issues more generally manifested in a number of topics—restrictions on seed saving, availability of conventional alternatives, risks of drift, and equipment and infrastructure changes—that all related to whether farmers' technological dependence would ultimately render them hostage to the whims of their single technology supplier. There was a sense that many of these issues were in flux and everyone concerned was poised to see the outcome. Competition was the solution on many lips, but often with some ambiguity as to whether this salvation was probable or just desirable. Interviewees were also asked about the lawsuits between Monsanto and farmers in the area. While many had opinions, and some hoped the outcome would positively affect their future, there was little sentiment of a commonality of fate or of solidarity: those who took the legal route took it alone.

Price Control

Very few agricultural stakeholders did not mention the technologies' high cost. These sentiments ranged from "it is very expensive but at this point it is still worth it" to various articulations of the opinion that it is hugely, unreasonably expensive and there is no end in sight. Prices had just taken a dramatic jump in 2005, and even among those few who accepted the cost as fair value, the majority qualified their position as dependent on there being no further such increases. Producers must purchase both the seed and the technology fee, and costs were on the increase for both:

> The costs are something that I think has got to be addressed. . . . For RR cotton it doubled, pretty much: it went from $13 to about $25 to $26. . . . That's for the tech fee. That's for their capped rate that is about 3.3 seeds per foot. If you aren't in their program of course it increases. (MS#19, agricultural expert, cotton)

As this quote indicates, a variety of factors affect the final price. These factors change according to the pricing scheme, which was modified in 2005. Changes in the pricing structure also make cost comparisons over time difficult. For example, in the RR system, the technology requires the application of a glyphosate-based herbicide, such as Monsanto's Roundup. Once Roundup was off-patent, it faced significant competition from generic glyphosate herbicides, and Monsanto made the use of Roundup a contractual provision of its TA for RR crops. This provision later had to be removed, as will be discussed in the context of the litigation. In an apparent response to the consequent loss of its herbicide market to competition from generics, Monsanto lowered the price of Roundup, but increased its technology fee. Similarly, while the technology fee was formerly charged on a per-acre basis, it was changed to a per-seed basis in cotton. Changes such as these are the root of two significant issues raised by stakeholders: the first, simply stated, is price control; the second relates to production control. That is, Monsanto's various rules and rewards programs increasingly dictate aspects of production that previously were a farmer's prerogative.

First, with respect to price control, the high level of frustration over the cost increases was irreversibly tied to the sentiment that farmers have no choice but to take it. The following characterize the sentiment quite well:

They pretty well dictated this year. They came in and increased their soybean fee, and increased their cotton fee, and increased their other fees, and what can you do about it? I mean as a farmer, and as a consultant, because we have to deal with their economic situations, what can you do when this company comes in here and says we're going up $20 a bag on this stuff and going up $5 a bag on this other stuff, and you're sitting up here saying, "I've got no other choice." (MS#29B, agricultural consultant/producer)

They are scaring us. If we do have a chance of making any money in the future, we're scared that they are going to keep on raising it. And we're going to get squeezed here and there, wherever we get to have some extra income. We're scared they're going to be raising it more and more. (MS#24, GM producer)

In justification of the cost increase, Monsanto responded thus:

Generally, our pricing philosophy for our trait technologies continues to be based on sharing the profit potential delivered growers by our products versus the cost and benefit of the alternative products that they may otherwise use. (MS#40, trait stewardship manager, Monsanto Company)

Those who thought the cost was still fair for the benefits they were getting were rare, however. The general sentiment from those interviewed was that now that 97% of farmers had bought into the GM system, they were trapped into paying whatever price was set. A particularly sore point with soybean farmers was the price of the seed, which they were no longer permitted to save due to the patents on RR soybeans.

The recent history of seed saving varies by crop. Seed-saving restrictions have had no impact on corn production, for example, as corn is a hybrid crop. Hybrids do not reproduce offspring genetically consistent with the parents, so producers must buy fresh seed every season. Hybridization itself was a social decision similar to patenting, but beyond the scope of the present discussion (see Kloppenburg, 2004, for more). Saving cottonseed is technically possible, but not commonly practiced. Cotton has to be ginned and then acid-delinted so that the seeds can fit into a modern seeder. The infrastructure for doing this has declined, and acid-delinting is only available in a limited number of facilities, mainly outside the state. While the opportunity for saving cottonseed is thus already practically curtailed by the decline in these facilities, few seemed interested in doing so in any case. In fact, many cotton and soybean farmers reported that they had not saved seed in the past because it was cheap and plentiful — quality-controlled seed could be easily purchased, without the hassle of seed saving. The sentiment "Why not buy good quality seed because it wasn't that expensive anyway" (MS#6) was common. As one farmer explained, even those farmers who would save seed typically still bought fresh seed every few years in order to get the newer varieties.

Restrictions on seed saving brought by GM technologies therefore mainly impacted soybean farmers, and for many of these it had only a very modest practical impact. This impact likely differs somewhat by size of producer, as smaller outfits are more likely to use seed saving to control costs. With GM technologies' contractual obligation on producers to purchase seed anew every year, not saving seed became dramatically more expensive, however. Now that GM soybean farmers had to purchase seed each year, its price climbed in response to the trapped market: specifically, it had "gone wild" to the point where soybean seeds worth only $6 on the commodity markets cost producers $30 to purchase as an input. Soybean producers were keenly aware of their limited options:

> We can't save our seed. They've got the variety and we think it's the best variety to plant. We have no choice, we've got to pay this exorbitant price for the seed. . . . They do a markup, but that's because they've got something we have to have, and we've no other way of getting it. (MS#34, GM producer)

Rising seed costs not only make farmers feel hostage to cost increases, but also affect their production practices. For example, high seed costs affect planting rates, as farmers try to reduce their input costs by planting less. In addition to price increases on soybean seeds, the change to pricing cottonseed on a per-seed basis — with the newly increased technology fee now included in the seed price — shifted the bulk of the high input cost of GM farming onto the seed itself. Consequently, whereas seed used to be the cheapest of a farmer's inputs, it quickly became something to regulate and minimize whenever possible. As one GM cotton farmer described this change:

> Not too many years ago [cotton seed] was around $50 a bag. You go back, it was really cheap. The seed was the cheapest thing. You could just plant a whole lot of seed. When I was young you'd plant a whole row of

seed, then you'd cross plant to get the right population, and then they'd block out, to cut down the population that way. But now you want to know exactly how many seed. You plant exactly the same. I was planting two seed every 8 inches. (MS#6, GM producer)

Precision planters now make it possible for farmers to regulate the exact depth, width, and placement of seeds, and so designate the exact number of seeds planted per acre. While reducing the seeding rate reduces costs, producers risk not getting the desired yield if they reduce it too far. Monsanto's profit strategies are not limited to seed and technology fee pricing, however, but include their rewards program, which is again linked to the seeding rate. An agricultural expert explains below how the change to per-seed pricing created a powerful incentive for farmers to join the rewards program:

If you use their program, then they cap your seed technology fee at $25 per acre in that instance. Whereas if you use a generic [chemical] . . . then you don't get that cap. So if you did plant more seeds per foot than the 3.3 [which make $25/acre] or whatever their cap is, then you are actually paying more for the technology. The way they've done it is really, I mean it's brilliant on their side, from their marketing program, if you don't use their product and you plant a higher seed rate then you actually pay a lot more for their technology. (MS#19, agricultural expert, cotton)

Admiration for Monsanto's marketing tactics aside, the narrowing of options for producers is apparent as "incentives" are considerably compelling. Planting a rate higher than Monsanto's cap (which the above expert emphasized was "not in our recommendation" as it was too low for desirable yields) is exorbitantly expensive without committing to Monsanto's rewards program, which is designed to maintain customer loyalty to the use of Roundup. By restructuring prices and organizing the rewards program in this way, the incentive for farmers to use Roundup over a generic is considerable.

Replant protection gives farmers another strong motivation to join the rewards program (and use Monsanto's herbicide), as farmers in the program who experience a planting failure are provided with a rebate on repurchasing seed. Given the difference in price between Roundup and generic herbicides, this still wasn't sufficient motivation for some. With the changes in pricing to include the technology fee in the seed (instead of charging per acre), however, the incentive greatly increased. Replant refunds appear to vary, but seed companies often offer farmers some sort of replant guarantee, where they would replace a portion of the seed, for example 50%. With the technology fee included in the seed, an affected farmer would have to repurchase it as well as the seed for his repurchase portion: thus if 50% of the seed must be repurchased, Monsanto is ultimately compensated 1.5 times for the technology fee.

Monopoly Control

While any company that is the sole proprietor of an essential product is going to be the recipient of some consumer objection, the negative perspective of Monsanto significantly outstrips this sort of grumbling. Farmers and stakeholders repeatedly indicated a negative relationship between producers and the company they are beholden to for their technology. The following provide good examples:

> They don't want just what I've got and he's got, they want what you've got—they want it aaaaaall! And they got greedy. It's been very obvious. In cottonseed, they are charging by the seed. . . . Who the hell are they kidding? . . . [Growers] hate them. You know, the little crap they keep coming along. They coming along, they want you to sign contracts, then they want to charge you by the seed for cotton, they just keep changing the rules again. They already got 98% of the market. Who gives a damn about signing contracts and then count your seed and put them in a bag! . . . They got the grower, they got him tied he can't go any other way and he hates that. He resents it. . . . They've just got so many little rules. (MS#20, agricultural expert, soy)

> I don't think you could find a farmer in this region that likes Monsanto. They have the worst PR that could possibly be. . . . It seems like they would be trying to work on the PR and do good things, but it just seems like the whole company is arrogant. . . . It's kind of like, "We're it, and you all have to go through us, and we don't care." (MS#24, GM producer)

Negative sentiments are not only expressed over cost and price increases, but frequently extend to issues of control and relative power. There was no doubt in the minds of those who dealt with the company that Monsanto was "it." While a few hesitated to call this an outright monopoly if state and federal officials declined to designate it as such, the majority had no such qualms. Monopoly in name or not, they had no doubt that Monsanto ruled the roost in Mississippi agriculture:

> Monsanto's got you where they want you. They've positioned you, as opposed to positioned themselves. (MS#26, GM producer)

> Even though you've got different bags with different names, it's still the same choice. And that one individual still has control over that choice. And he's going to get the same amount of money no matter what he does. (MS#29A, agricultural consultant/producer)

In short, there is a general sentiment that Monsanto makes the rules for the agricultural community and that these rules shift according to Monsanto's strategies for maximum benefit. There is a distinct sense that now that farmers have become dependent on GM technologies, they are subject to Monsanto's whim. This awareness of power imbalance even affects the seed dealers who have contracts with Monsanto. For example, one dealer shared his concern that as GM competitors came on the market, he would be forced to gamble on which he thought would be the winning technology, as Monsanto could pressure dealers to maintain a certain share of their business or forfeit their license. This would, in turn, affect a producer's options.

Thus GM producers are facing a clear reduction of control as a result of their adoption of biotechnologies. At the same time, farming in Mississippi has changed to such a great degree as a result of this adoption that many have become almost dependent on GM technologies. Within ten short years since the first GM introduction, it is not in any way unusual or striking to hear a farmer make statements such as the following:

> We can't grow cotton without the Bt. (MS#4, GM producer)

> They're holding all the cards; what are we going to do about it. They're like, "Don't buy our seed and go back to conventional cotton," but no, we can't do that. (MS#6, GM producer)

If this dependence is indeed the current state of agriculture in Mississippi, the challenge to producer control would seem significant. Of course, the resounding response to such implications by the technology's staunchest defenders is that producers can simply go back to farming conventional seeds. This begs the question of whether farmers could indeed opt out of biotechnologies if they so desired.

A Return to Conventional Agriculture?

Interestingly, the availability of conventional seeds is an ambiguous point, with arguments on either side. On one side are those who argue that conventional seeds are available, even good-quality ones, going so far as to show me seed listings and their yield reports. On the other side are those who argue that conventional seeds are incredibly difficult to obtain and aren't available in good yielding varieties. There are some plausible explanations for this polarity. For example, there are ample indications that some institutional agents want to protect against negative perspectives of the technology, which assertions of a lack of choice might support. On the other side of things, some producers might try to soften the contradiction between their expressions of outrage and their continued use of GM technologies. Nonetheless, strict availability aside, there are clear signs that a retreat to conventional seeds is quickly losing viability. The fallout of the 2005 price increase provided an excellent example of this.

The 2005 fee increases were substantial enough that many angry and frustrated farmers talked about reverting to conventional crops. The sentiment seemed greater in the hills area, where farms are smaller, drift susceptibility is lower, and resentment over the technology is exacerbated by the impact of a cross-state cost differential based on the higher profitability of the delta region. In any case, enough interest in conventional seeds was generated to create a supply crisis. A long-term conventional cotton farmer described his experience:

> This past year they increased the RR fee and the Bt fees quite substantially. Enough to really make people around here really rethink buying some of those cotton varieties . . . [An acquaintance] got the same seed dealer I did to search for conventional cottonseed, and got brought in conventional cottonseed and planted it on their land for the first time in years, this year. Because they were highly upset about the tech fee increases, and so more or less as their way of rebelling against that, they chose to go back conventional. As a matter of fact there were several farmers looking into doing that and the seed availability just wasn't there. Thankfully we got in on the front end of it and got our seed. A lot of

people were still looking for seed right up to planting time. (MS#12, conventional producer)

Similarly, an expert in soybean production explained how farmers upset by soybean price increases and likewise interested in abandoning GM varieties in favor of conventional ones were thwarted in their efforts:

I had more calls this year from farmers wanting to grow conventional varieties, with farmers whining and moaning, "I can't get them; what am I going to do?" (MS#20, agricultural expert, soy)

Availability, of course, relates not just to strict availability, but to the availability of good-quality, high-yielding seeds. Even where conventional seeds are strictly available, as in Saskatchewan, many claim that the varieties are not good. Companies are phasing out conventional varieties, and the vast majority of new varieties are only offered in RR, Bt, or RR/Bt stacked. As one GM cotton producer describes the availability issue:

I started to plant some conventional cotton, and there are almost no . . . Out of, I don't know how many varieties are available on the market. There might be 200 varieties available. You can probably count every [conventional] variety available on one hand. And probably not need every finger to do that . . . Well, when I say not having choices available, the choices of good yielding varieties. There's not but about three available, and one that would have a yield comparable to the genetically modified crops, and there was very little of it available. (MS#14, GM producer)

There is consensus that if farmers wanted to revert to conventional seeds, there isn't the supply available for them to do so, at least not in the immediate sense. This lack of availability is partly a result of low demand. Farmers wholeheartedly embraced GM technologies, and in seeming consequence of the laws of supply and demand, conventional seeds fell out of production. Quality debates aside, seed dealers can still obtain conventional seeds on the market for the few who want them and have them shipped in. The problems arise when the need is immediate and full scale, given the time it takes to produce seed. As one seed dealer explained, his soybean production was currently 100% RR, and if there were an unexpected shift in demand he would not be able to supply his customers without at least 14 months' notice. If he then had enough demand for conventionals, he could either produce them or contract someone else to do so. The agricultural agent who reported on the complaints he received about the lack of conventional soybeans further supports this perspective:

> You know, when you go less than 10 years, going from 100% conventional to 98% Roundup, it's a big change. You can't expect someone to grow you conventional on the whim that you might buy. . . . We had growers who wanted to grow conventional and couldn't get them, couldn't get the seed. And they had a hard time understanding that. And they probably will continue to have a hard time, but it's not a decision you make overnight. You can't decide today, I'm going to have conventional varieties for next year. That decision is for two years down the road. (MS#20, agricultural expert, soy)

Yet, despite the apparent demand for conventional seeds, the dealer I spoke to was still looking at growing only a small fraction of them, if any. In part, this might have been because, as claimed by some, while farmers will say they refuse to spend $30 per bag on soybeans, in the end they will because of the relative hardship of reverting to conventional production. If so, further cost increases could steel resolve. Meanwhile, there has been a significant retrenchment of public breeding programs, which operate outside of the profit system. While there is still some public breeding in existence, as in Saskatchewan, conventional varieties are not released anymore. Without public breeding, farmers not only lose a source for conventional seed, should they ultimately decide to revert to conventionals, but they also don't have alternative, publicly bred varieties to act as a check on the price demands of private companies.

Technological dependence is not just a function of seed availability, however. A number of structural changes associated with GMOs also make a return to conventional production difficult. Infrastructure problems, for example, are already evident for the few cotton producers who farm conventionally, and would like to gin and acid-delint their seed. The difficulties in cleaning and storage are furthered by liability concerns, which make seed cleaners reluctant to handle conventional seed in this way. Available equipment creates another dependence. Just as GM technology has reduced cultivation and chemical applications, so it has reduced producers' equipment needs: those who no longer have sufficient working equipment will be unable to return to conventional production without significant capital input. For those who upsized, expanding their acreage in response to the reduced labor requirements of biotechnologies, these new acres themselves become an impediment to reverting to conventional production. Drift, as noted, is another issue that can hinder reverting to conventional crops, as without a full-scale shift, using non-RR crops entails a significant degree of risk.

For many producers, their current relationship with the makers of the technology—and their relative disadvantage—is enough to concern them. For those who considered recent changes in the context of their future production, control issues took on even greater salience. For many of these, economic concerns about being beholden to one company became further linked to broader concerns that when a problem develops, this one company may not be ready to address it. For example, while those in agriculture expect weed and insect resistance problems, their dependence on one company for the solution to these problems is a new development that renders them vulnerable. What for Monsanto could be a lost market opportunity—for example, the failure to come out with a new product in time when resistance develops—could spell economic devastation for farmers. An agricultural consultant who was adamantly anti-monopoly and who harbored no doubts about Monsanto's status in this regard, articulated the concern well:

> I do not like the idea of Monsanto owning everything in the country, and us having to rely on one company to furnish us with what we have. I am more anti-Monsanto than I am anti-GMO. I consider GMOs a tool. How long that tool is going to be available and effective, I don't know. . . . What concerns me is that because of the GMO-Monsanto relationship . . . we've destroyed the ag-chemical business. . . . You know, I don't have researchers I can call today that I could call 10 years ago and ask questions, because they don't exist, they don't have jobs anymore. The marketplace is not there. Monsanto owns the marketplace. . . . It concerns me what is going to happen five years from now. As the mare's tail becomes resistant . . . as other things become resistant—what is going to be there, and is Monsanto going to be the only person that's going to be available to furnish us with what is there because the others are out of business? (MS#29A, agricultural consultant/producer)

For those who have become completely dependent on the technology—through lack of equipment, greater acreage, or loss of skills—a market failure by Monsanto along these lines could be the end of their business. The issue of alternatives is therefore both an issue of control and an issue of having something to go back to in the event of a failure of the technology.

Fields of Opposition?

While Mississippi farmers clearly love GM technologies, the open resentment of Monsanto and the host of control concerns would seem likely to elicit some active opposition. Despite the highly negative attitude toward the company and the fulsome complaints about cost and control issues, however, actual struggle against Monsanto appears negligible. Certainly, resistance is loudly promised—and perhaps even seriously considered— for example, by those who thought of reverting to conventional varieties after the price increases. Outside of this, however, there were no signs of lobbying, subversive seed saving, or any other forms of active resistance.

Two factors seem to figure largely in this lack of resistance. First, despite all the control concerns, producers still appear to have a great deal of faith in the workings of the market. A frequent mantra of those interviewed was that with competition, the control issues around GM technologies would be resolved. Thoughts on their frustration with Monsanto would often be completed with statements about the ameliorative power of competition. In part, this faith in competition was likely bolstered by an abhorrence of regulation, as well as by the promise of new GM technologies on the horizon.

Second, producers expressed some fear that they could lose the technology which they have come to depend on. More than once farmers expressed the sentiment that they didn't want to "rock the boat" so much that they risked losing the technology. One GM producer who considered returning to conventional crops after what he called Monsanto's "price gouging," for example, ultimately decided against it because of his fear of operating without the technology:

> I guess I've also gotten somewhat addicted to or dependent on the technology and wasn't completely ready to let it go. Maybe a little bit of me was . . . I used it so long it's like someone on crutches. I've used it so long and you're kind of unsure—it's like training wheels on a bicycle, you're not ready to throw them off quite yet. You could probably do it, but you're not 100% sure of going back to doing some of the older ways. (MS#14, GM producer)

Such outright expressions of technological dependence were rare, but the sentiment was not. Producers were not shy to state such an affinity for the technology that it smacked of dependence:

> I really think it's [Bt] a necessity. We went back to conventional, it would be tough. (MS#3, GM producer)

This dependence was much greater for Bt, given the concerns over worm control, but it was evident for the ease, convenience, and efficiency of Roundup as well:

> I think that when they see the prices they will respond to it, but when you hold them down to the line, they are going to plant it every time because it's a lot easier. (MS#19, agricultural expert, cotton)

> It's so much easier raising a crop now than it was 10 years ago. So I don't want to go back to the way we were. (MS#24, GM producer)

While no resistance appears to have been initiated from within the agricultural community, efforts have been made from the outside. For example, a number of producers reported that some attorneys from Atlanta, Georgia, attempted to initiate a class action lawsuit against Monsanto in an effort to recoup some of the technology fees on soybeans. Even then, resistance to Monsanto's control was balanced against producers' fear of losing the technologies' benefits. As described by one farmer who went to a meeting on the issue:

> Every one of us in that room and everyone that was at the meeting knew that we paid x number of dollars, and we knew what we paid it for, and rather than take the chance of not having the availability of the technology. . . . I'm not willing to take a chance and lose something that's good. (MS#31, GM producer)

Producers' relationship with GM technologies therefore appears to operate in a tension: on the one hand, their appreciation of the technology has led them to increasing dependence on it; on the other hand, they are experiencing the drawbacks of this dependence in their relative lack of power with respect to its makers. Without a doubt, these drawbacks (particularly with respect to control issues) will only increase as the technologies become entrenched in more crops. While the trajectory of negative outcomes resulting from agricultural biotechnologies is fairly clear, at this point the benefits are still too great for producers to offer any significant resistance. This dynamic is captured in the accounts of two producers who themselves tried to explain the lack of resistance:

> If I'm raising RR cotton, and RR soybeans, and you know, my stress level is a whole lot less, and come Friday I'm through with my work and I don't have to worry about working all weekend, and I'm making a living, then I'm happy. Until it gets to that point when it's strangling somebody . . . I just think [Monsanto's] britches are going to get too big for them one of these days, and [by then] there's just not going to be enough competition left. (MS#29A, agricultural consultant/producer)

> Everybody is concerned but they all want the ease of using the technology. Don't worry about it, they'll just take the ease of using the technology. So it's the kind of thing you're always going to do later. We'll always do it later. We'll look at this contract closer and find loopholes in it. Well, it just never happens. (MS#10, GM producer)

Environmental issues could be another trigger for resistance to the industry, but the environmental and health issues that act as a lightning rod for biotechnology-related controversy outside of the state do not appear to be a significant issue within it. As we saw in Saskatchewan, the economic imperative of modern agriculture is usually so predominant that support for any potentially beneficial technology is unequivocal from those within it. Opposition is most likely to come from those outside industrial agriculture, such as organic producers and environmental organizations. In Mississippi, these outsiders were hard to find.

In 2002, the first year for which data on certified organic farms was available, Mississippi had 160 certified organic farms out of 42,200 total farms (USDA, ERS, "State Fact Sheets"). Thus less than half a percent of all Mississippi farms are certified organic. Prior to 2005, there was no organic certification board within the state. The only related grower support organization—the Mississippi Organic Growers Association—was defunct by 2005, and the past organizer stated that he just couldn't get enough members. Those I interviewed who were more environmentally oriented reported that many small producers farm sustainably, but certification involves too much paperwork. Therefore, despite the lack of specifically organic organizations, there are a few grassroots organizations promoting sustainable agriculture—broadly conceived as environmental as well as economic sustainability—many of which are concerned with low-resource producers. Two such groups are the Sustainable Agriculture Working Group, which focuses on sustainable agriculture, particularly through promoting small producers growing for farmers' markets, and Mississippians Engaged for a Greener Agriculture, which promotes sustainable farming primarily among the African American farmers of the delta. Environmentalist attitudes can also be found in the Sierra Club and the Green Party.

Members of these groups are more likely to raise objections to GM technologies. Perhaps due to their limited numbers, however, even their objections are partnered with a perspective that the continuation of the technology is inescapable. Organized resistance to biotechnologies on health or environmental grounds is virtually nonexistent in Mississippi, and all queries, wherever directed, were met in the negative:

> We really haven't had any issue with Frankenfoods or anything like that. . . . Whether it's safe to eat or to grow. It's really not an issue here, whether it's bad for you. (MS#3, GM producer)

As a member of the newsprint media explained to me:

> The people who care about something of that nature [GMOs] are typically in college towns. Mississippi has college towns that have a very diverse culture, and the rest of Mississippi is not really diverse at all, and those would still be new ideas to those towns. (MS#23, media/organic producer)

Even in the more environmentally progressive groups, objection to GM technologies cannot be assumed. Despite the anti-GMO position taken by the national level of the Sierra Club, for example, the Mississippi chapter does not advocate this position. With respect to sustainable agriculture producers and organizations, there appears to be some distancing from GM technology issues, likely in part because the vast majority of sustainable producers are involved in fruit and vegetable—rather than row crop—production. Some concerns were raised regarding the availability of non-GMO seeds, however, given Monsanto's ongoing seed company purchases.

> [If Monsanto] continues that, it will get to the point where there would no longer be organic seeds. I don't see where he would have a capture

there, or a market there, for organic seeds at the same time as he is push-
ing or promoting all the different types of herbicides and pesticides.
(MS#28, organic producer and sustainable agriculture activist)

The lack of environmental concern around biotechnologies appears
consistent with the environmental values of the state more broadly. Whatever their differences, members of the Agricultural Producers Association
of Saskatchewan are not likely to refer to organic farmers as "nutcases,"
as did a member of the Mississippi Farm Bureau, the largest general farm
organization in Mississippi. This value difference is not a limited personal
view but found frequent expression, for example in the agricultural expert
who argued in favor of the high-capital input of GMO farming as "the
right way to go," but against organic farming for disadvantaging the commodity prices of those who "can't afford" to take that route.

The environmental picture that emerges in Mississippi is thus very
different from that in Saskatchewan. Where strong resistance to GM
technologies can be readily found in Saskatchewan's alternative agriculture sector, it is largely absent in Mississippi. Further, what little objection there is to biotechnologies in Mississippi on environmental grounds
operates with little apparent connection to—or even much awareness
of—the control issues raised by farmers. This is again in stark contrast to
the extensive networking of anti-GM alliances in Canada. While resistance to the social reorganization brought by agricultural biotechnologies
in Mississippi is not forthcoming from environmentalists or from either
conventional or sustainable agriculture producers, it is evident in the very
limited forum of the patent infringement lawsuits initiated by Monsanto.

Mississippi Farmers in Court

Everyone I spoke to had heard, in general, about Monsanto's lawsuits
with certain farmers in Mississippi over seed saving. While the details
and significance of the cases will be discussed further in the following
chapter, the current discussion regards what role, if any, they played in
a reorganization of agriculture. Overall, farmers were well aware of the
issues of seed saving and antitrust raised in the suits. Often the awareness appeared to go no further than this. To some extent, this appearance
might have been the result of a southern gentlemanliness at play, which
precludes naming names and discussing what might be the equivalent of
"dishing dirt" on private affairs. One regional agricultural representative

suggested for this very reason that it might be best if I stick to generalities rather than mention names when interviewing people. This contention appeared to be borne out in a number of interviews, where interviewees would refer to vague knowledge of a gentleman engaged in such a lawsuit, while further discussion often revealed their knowledge to be far greater.

Generally speaking, there was a fair amount of empathy for the issues raised by the farmers in the lawsuits. Nonetheless, two clear categories of perspectives emerged: those who subscribed to a clear view of right and wrong and regarded seed-saving violators accordingly, and those who had broader concerns with the balance of power between companies and farmers, and who consequently had some empathy for the farmers involved, regardless of their guilt. While distaste for those who blatantly violated contracts was palpable, the latter group nonetheless greatly outnumbered the former.

In the first group, strong views of right and wrong precluded any consideration of the circumstances that might have led to the seed-saving violation. For these respondents, farmers who signed contracts knew what they were agreeing to, and if they saved the seeds anyway, they knew the risk they were taking. The following exemplify the perspective:

He knew up front that he wasn't supposed to do it, and he did it anyway for his own profit. In that sense in itself, you know, I think he was wrong. I don't agree with the amount they charge for the technology fees and the seed and whatnot, but if I have an alternative and I don't use it, then it's my choice. (MS#14, GM producer)

I think generally, most folks know, if you sign a contract, you sign a contract. It's an agreement between two parties. If it says I will not save seed for replanting in honor of the agreement, I think generally, most everybody is going to honor that side of it. (MS#5, Farm Bureau)

I'm on the side of Monsanto in that case. We are using something that they developed that they are selling and they are going to continue to develop things in the future that we need if we are going to stay in business, and they need to have that return. If farmers don't want to pay the fee, they can plant conventional varieties. (MS#3, GM producer)

There was little empathy for the farmers involved, who were seen as dishonestly attempting to profit from what was commonly known was prohibited. These interviewees felt that the technologies provided a benefit, and that laws were in place to ensure that those who developed them captured returns sufficient to warrant developing new technologies. As the above quotes indicate, a key contention of many who held this position was that those who didn't like the rules should simply opt out of using GM technologies.

For those who were more concerned over what could be characterized as balance-of-power issues, such concerns overrode considerations of the specific infractions involved. These respondents mainly focused on the issues of cost and monopoly power, and frequently referred to the lawsuits with a somewhat passive "I hope it helps" sentiment.

> As far as my viewpoints, as far as me wishing they'd stick it to [Scruggs] or nothing, I don't. I hope he gets the point across to [Monsanto]. I'm hoping he does something that will help us. I hope they realize that they are going to have to do something with us to help us turn a profit. (MS#13, GM producer)

Farmers with this perspective were also the most likely to be concerned about diminishing alternatives to GM seeds, whether due to a lack of availability of conventional seeds, their dependence on the technology, or other reasons.

Most commonly, such farmers balanced what they saw as a distasteful infraction with an acknowledgment of broader issues that perhaps could only be settled through this type of action. This balance is perhaps best captured by one of the few producers who ultimately opted to grow conventional cotton, although continued to grow GM soybeans:

> The one thing about the lawsuit—and I won't even mention names, everybody supposed to know who it is—but they said that the man that's involved . . . the big man that they are really after, he is the only man that is strong enough to fight Monsanto and possibly win. So they said he has the best potential of anybody of knocking them down. So everybody is kind of rooting for him in that respect. However, nobody agrees with what he did that I have talked to. They know why he did it. They know why he was mad enough to do it. The way I understand it, he saved some seed when he wasn't supposed to. He claims they are his seed. He bought and paid for them, he can do whatever he pleases. We understand where he is coming from, but we all knew what the contract said. (MS#12, conventional producer)

Essentially, there is a sense that breaking a contract is tantamount to being ill-mannered: it is not a southern thing to do. At the same time, concerns about Monsanto's monopolistic character and the antitrust manner in which it conducts its business—although not always articulated as such—prevail. For the most part, these perspectives are held in loose association. While there is not a great deal of affinity for those who engaged in unlawful seed saving, there is a mutuality of concern over Monsanto's seemingly inappropriate level of control and an acknowledgment that going to court might be the only means of addressing it.

Outside of the issue of seed saving and farmers' positioning on the broader control issues raised by the lawsuits, there is no evidence that the litigation affected the actions of Mississippi farmers. No doubt, the extent of GM adoption renders contamination concerns moot for most farmers. It is telling that the only interviewee to raise the issue of a potential personal impact from litigation was also the only farmer who grew conventional cotton:

A farmer my size, I couldn't even pretend to go to court to fight them. You know, if they said I did something wrong, and they want me to pay, I guess I have to, because I haven't got $20,000 to go to court to fight them, even if I know I'm right. Because it wouldn't be worth it to me, depending on the penalty. (MS#12, conventional producer)

Nevertheless, as in Saskatchewan, it is clear that Mississippi farmers' primary motivation for adopting GM technologies is their sincere appreciation of it, rather than fears over liability.

Even though respondents generally disapproved of the farmers involved in the lawsuits, they articulated the idea that, barring regulatory intervention at the state or federal level, lawsuits appear to provide the only avenue for ameliorating the power imbalance between Monsanto and producers. For example:

Without a challenge in the courts, and the laws being changed, we will not go back there to farmer control. Why should they give it up? Why should the companies give up something they've already got? (MS#20, agricultural expert, soy)

Conclusions

Within each individual farmer's survival strategy, GM technologies make sense. At current prices, depending on crop, they are arguably on economic par with conventional varieties while providing some significant advantages over them. With respect to Bt, the benefit is risk reduction—not a minor consideration in light of the 1995 heliosis outbreak. For RR crops, efficiency and convenience are the bigger factors. What little objections to GM technologies there are in Mississippi do not share the characteristics of those in more environmentally reactive areas such as California: the few who acknowledge any associated environmental or health controversies are primarily concerned with how they could impact the marketing of their product.

While individual farmers are rationally choosing to adopt GM technologies, these individual decisions are manifesting in a structural reduction of their choices—something that is only likely to increase as more GM technologies are introduced. As long as the technology is working and priced just within reach, the drawbacks to such a structural shift may not be readily apparent. They are not likely to remain concealed for long, however, as biotechnologies' form of dissemination has afforded vastly greater power to their maker, rather than their users. The extent to which this agricultural reorganization entails a form of expropriationism is highly dependent on farmers' ability to opt out if the conditions for using GM technologies become too unfavorable.

While still listed and technically available, conventional cotton and soybean seeds are definitely not as abundant or readily obtainable as GM seeds. Further, many felt that the available varieties were not as good, causing significant yield loss. Some stated that if they could purchase non-GM crops that could provide the same yields as GM ones, they would. While the actuality of this might be quite different, the rising prices definitely intensified interest in doing so. Drift was another issue. To the extent that producers feel they must purchase GM technologies to forestall such problems, expropriationism becomes much more salient. A further question regards the structural viability of a return to conventional production. Biotechnologies help farmers manage larger tracts of land in a business where profits are based on volume. Those who expanded would have the most difficulty reverting to conventionals. The retrenchment of the infrastructure for saving cottonseed, as well as the chemical and equipment base for conventional farming, also hampers a return to conventionals. In the context of competition, these developments could be seen as the inevitable drawbacks of any technological evolution. In the context of proprietary seeds and a single supplier of the necessary technology, however, this technological dependence becomes highly significant to the social structure of farming.

The sense from Mississippi farmers is that they have lost control. While this loss is largely framed in terms of cost and lack of alternatives, it includes all the control issues evident in the restrictions on seed saving, differential regional pricing, and numerous other changes instigated by Monsanto, in everything from its pricing schemes (per acre vs. per seed) to its incentive and rewards programs. In the manner of death by a thousand cuts, this can be characterized as loss of control by a thousand schemes. While most stakeholders would likely accept a technical determination that Monsanto does not have a monopoly, the majority I spoke with nonetheless convey that, in all practicalities, King Monsanto is ruling strong.

Resistance to GM technologies is nonetheless virtually nonexistent, though talk was high after the 2005 price increases. Some expressed faith that competition would save the day. Given the drift issue, the shift to another GM product in the delta would face a double hurdle of market competition and drift risk to early adopters. Moreover, with the corporate benefits gained by restructuring agriculture through biotechnologies, even where limited competition arose, it could still choose to replicate, rather than compete with, the imbalanced patterns set by Monsanto.

Expropriationism is therefore clearly on the advance in Mississippi. There is no doubt that significant restructuring is occurring as a result of GM technologies, largely contrary to the interests of farmers. Numerous variables could affect this trend, of course. However qualified, competition is one such variable. Regulation is another. Mississippi producers clearly prefer to avoid regulation, as evidenced by the drift issue. Consequently, many interviewees stated that no new regulations were required, outside of educating other people and nations not to reject GMOs, although some suggested that given their dependence, "something" was required to ameliorate the power imbalance between Monsanto and producers, primarily with respect to prices. While regulation is anathema to producers, few other means of adjusting this imbalance seem available. I will now turn to a closer investigation of *Monsanto v. Scruggs* and *Monsanto v. McFarling* to enquire into how farmers' control over agricultural production has fared in the legal forum.

CHAPTER 7

Starting a New Regime: Training the Locals

Farmers still have control over what seed they plant and whom they purchase it from. Restrictions on patented technology [do] not take away farmers' choice.
MS#40, TRAIT STEWARDSHIP MANAGER, MONSANTO COMPANY

A New Legal Framework for Agricultural Production

In the preceding chapter, we saw that the motivation for adopting GM technologies in Mississippi is strong. Whether due to the technologies' merits, the risk of chemical drift, or the perceived relative disadvantage of competing with those who do adopt, there are very few who choose not to use GM cotton, soybeans, and, increasingly, corn. Soon after introduction, Monsanto's GM technologies became the basis of agricultural production in these Mississippi crops. Of course, the consequent changes to Mississippi agriculture do not just result from the technologies' physical attributes, or even from the difficulties of dealing with an effective monopoly, but also from an associated legal framework that stretches from international and federal law down to the contractual provisions of Monsanto's Technology Agreement [TA]. To what extent is this legal structure reorganizing agricultural production in the United States?

This chapter asks the same questions as were asked of the lawsuits in Saskatchewan. As in that chapter, I will first briefly discuss the U.S. legal framework for intellectual property protection of plants, developed through legislation and case law around the patentability of life. I will then draw on the two court cases—*Monsanto Co. v. McFarling* and *Monsanto Co. v. Scruggs*—for specific consideration of the evolving legal framework around agricultural biotechnologies. As in Saskatchewan, many legal changes are evolving below the surface, their nature and extent—as well as their practical application—still unfolding. Looking at the key issues that arise in relevant litigation provides another means to assess these changes. I will then discuss resistance and draw some conclusions regarding the nature and extent of expropriationism revealed in the court cases, and what they suggest for its future trajectory.

The data for this chapter is made up of the decision documents from the two court cases, supplementary legal documents, and interviews with litigants and their representatives, where possible. As the number of such lawsuits in the United States far exceeds that in Canada, there is significantly more supplementary information available on the issues that arise and many more farmers involved. Given the presence of a case of some notoriety just across the border in Tennessee, interviews with two litigants there were also conducted.

Intellectual Property Protection

As we saw earlier, the United States strongly supports intellectual property rights protection, both nationally and internationally. Intellectual property protection for plant genetic resources in the United States was first granted in 1930 through the Plant Protection Act [PPA]. While the PPA offers an actual patent on plants, it is only applicable to asexually produced plants, such as ornamental plants and fruit trees (Evenson, 2000:14). Subsequently, the Plant Variety Protection Act [PVP Act], governing sexually produced plants, was introduced in 1970, a full 20 years before similar legislation was introduced in Canada. Unlike the PPA, the PVP Act did not offer patent protection, but rather a certificate granted by the USDA which provided a number of exclusive rights to its holder—such as rights to propagate, import, or export the variety. Broad exemptions to the PVP Act existed, however, such as allowing researchers to use varieties for breeding, and farmers to save seeds for their own use, as long as the amount saved did not exceed that originally purchased (ibid.: 16).

Since the PPA and the PVP Act, a third system of intellectual property protection over plants in the United States has emerged, in conjunction with the country's evolving membership in the International Union for the Protection of New Varieties of Plants [UPOV]. The United States became a member of UPOV in 1981, and in 1999 it became party to the newest version of the act, UPOV 1991. As discussed in Chapter 5, this version supports protection for breeders while rendering the exemptions for farmers and researchers optional. Under the 1991 version, the United States can either offer straight patent protection on plants or can continue with some sui generis system of plant breeders' rights, such as that offered by the PVP Act. Straight patent protection is, of course, much broader and does not allow for farmer and researcher exemptions; thus it would effectively end the practice of seed saving for any patented varieties.

Even before becoming a signatory to the latest version of UPOV, there was strong evidence of a U.S. predisposition toward plant intellectual property rights protection. According to Roberts (1999), the country has been "clearly disposed to accommodate the vital new age of biotechnology," with both the U.S. Patent and Trademark Office [PTO] and U.S. court decisions demonstrating a long history of support for the patentability of living organisms (23). Vaver (2004) similarly argues that the U.S. congressional reports which accompanied the enactment of the 1952 U.S. Patent Act, and which claimed the patentability of "anything under the sun that is made by man," characterizes the tone of the U.S. courts and the PTO for the last two decades (2004:158). It is in these decades that sexually reproduced plants were clearly affirmed to be patentable under utility patents. Therefore, unlike in Canada, with its reluctant Parliament and mixed court rulings, the U.S. decision process on the patentability of living organisms has been considerably quicker and far less ambiguous. The American tendency has been for "instant patent gratification: issue first and ask subject-matter questions later" (159).

As in Canada, patent protection is a federal matter in the United States, is granted for 20 years, and requires that a process or product be novel, useful, and non-obvious. A patent can be obtained by anyone who "invents or discovers any new and useful process, machine, manufacture, or composition of matter, or any new and useful improvement thereof" (United States Code, Title 35, Section 101; "Patent Act" hereafter). American case law about the patentability of life forms found early roots in debates in the late 1800s and early 1900s over whether mere discoveries (rather than inventions) and products of nature could be patentable (for examples, see Roberts, 1999:23–24). Not until 1980, however, in the landmark case of *Diamond v. Chakrabarty*; *Chakrabarty* hereafter), did the patentability of living organisms find explicit judicial support.[1]

Chakrabarty involved a bacterial culture that could break down crude oil, useful with respect to oil spills. The U.S. PTO and the board of appeals had rejected the culture's patentability on the basis that it was a product of nature and that, as a living thing, it was not patentable subject matter (U.S. Congress, Office of Technology Assessment [OTA], 1989:8). Ultimately the U.S. Supreme Court decided that the human-made micro-organism was patentable as it was human altered. Notably, the form of human agency supported by the court excluded the collective, anonymous, and incremental agency that is involved in more traditional breeding (Aoki, 2008:42). In consequence of the ruling, *Chakrabarty* became "the linchpin to the explosion of biotechnology patents in the late 1980's and 1990's" (Dhar and Foltz, 2007:4)—the "stimulus to the growth of the biotechnology industry" (U.S. Congress, OTA, 1989:8), and the launcher of "a new paradigm in biotechnology patenting" (Dhar and Foltz, 2007:1). It remained to be seen whether such patents could apply to multicellular organisms, however.

Post-*Chakrabarty*, there remained some confusion over the issue of the patentability of plants in the context of the PPA and the PVP Act. This was resolved a short time later, in 1985, through the U.S. PTO board of appeals application of *Chakrabarty* to plants in the case of *Ex Parte Hibberd* (*Hibberd* hereafter).[2] Contention over the fact that granting utility patents on plants seemed to be in direct conflict with the intentions of Congress was further aired in *Hibberd*: why would Congress have enacted the PVP Act if they had not specifically wanted provisions that recognized the extraordinary aspects of intellectual property protection for plants? The question is highly significant, as granting plant patents would appear to negate Congress' intention to protect farmers' right to save their seeds. In *Hibberd*, the PTO appeals board ruled that the PPA and the PVP Act did not prevent utility patent protection for plants, and in the context of "congressional silence," an "administrative agency . . . made a very important policy decision: *plants are eligible for utility patents as well as PVP certificates*" (Aoki, 2008:43).

Granting utility patents on plants represents a significant loss to an ownership right—ownership of the progeny of their crops—that farmers held, and that had been affirmed under the PVP Act just 15 years prior. The contrast to Canada is clear, and further elucidated with respect to the Harvard mouse: while the mouse was rejected for patent protection by the Canadian Supreme Court in 2002, the U.S. PTO granted the same mouse patent protection as early as 1988. The significance of the change finally faced direct court challenge in 2001—"1,800 granted utility patents later" (ibid.:46)—in the Supreme Court case *J.E.M. Ag Supply, Inc. v. Pioneer*

Hi-Bred Int'l, Inc. (hereafter *J.E.M. Ag Supply*).[3] While the litigation involved an agricultural retailer, not a farmer, the case raised the question of whether the PVP Act "pre-empted the ability of the Patent office to grant patent protection for plant varieties" (Hamilton, 2005:52). Both the previous district and appellate courts upheld the validity of the patents, and, ultimately, so did the U.S. Supreme Court ruling. This ruling showed that the court was "not going to revisit the larger issue of the wisdom or legality of granting patents on living materials" (ibid.), but was "squarely holding for utility patents on plants" (Kershen, 2004:575). It would seem that in the clash of rights over patenting seeds in the United States, biotechnology developers had emerged the clear winners: expropriationism seems unambiguous.

In consequence of this case law, U.S. intellectual property protection over plants is currently provided by three systems: the PPA, the PVP Act, and utility patents, which can be obtained on both sexually and asexually reproduced plants. While utility patents offer the broadest protection, they have stricter requirements, and thus all three systems remain available to seed breeders. It is the utility patents, however, which have triggered the current struggle over the commodification of the seed. As we shall see, the conflict between utility patents, the PVP Act, and the right of farmers to save their seed found a second wind in Monsanto's patent infringement litigation.

The Lawsuits

While infringement lawsuits were possible under the PVP Act and the PPA, only a limited number were ever prosecuted (Kershen, 2004:575). With the patenting of GM traits under utility patents, that number was predicted to go up significantly (ibid.). By all indications, this is already well in process. In 2005, the Center for Food Safety [CFS] published a report on U.S. litigation over agricultural biotechnologies initiated by the Monsanto Company. The CFS estimated that Monsanto conducts 500 investigations of farmers in the country a year (CFS, 2005). It further extrapolated from Monsanto's own reports to conclude that the company had handled between 2,391 and 4,531 "seed piracy matters" in the United States by 2006 (CFS, 2007). By 2004, it had filed 90 cases against 147 farmers and 39 businesses/farm companies—either over technology use agreement violations or patent infringement—from 25 different states (CFS, 2005:31). By 2007, this number had climbed to 112 (CFS, 2007).

The majority of these cases are settled out of court for undisclosed sums, the details of which are protected by confidentiality agreements. For those cases that do proceed through the court system, it is clear that the economic impact on farmers can be considerable. Recorded judgments (which are often far below the actual cost of the entire proceeding for the farmer) range from a low of just over $5,000 to a high of over $3 million (CFS, 2005:34). The mean settlement for those with recorded judgments is $412,259.54 (ibid.). Given the high cost of litigation and the imbalance of power and resources, the vast majority of farmers — including some of the more outraged — appear to settle with Monsanto. For example, one U.S. case that began to garner some of the same attention as the Schmeiser case was Monsanto's suit against Nelson Farm in North Dakota in 2000. Despite the Nelsons' full cooperation and the fact that the State of North Dakota Seed Arbitration Board found no support for the company's claims,[4] Monsanto persisted in its allegations. The outraged Nelsons launched a website and initiated a publicity campaign, vowing to fight "on the principles of our innocence" (NelsonFarm.Net; see also Schubert, 2001; Witte, 2001). Monsanto dropped its case in 2001 in conjunction with an out-of-court settlement with the Nelsons, the details of which are protected by a nondisclosure clause.

Two court cases where farmers have not settled and that have risen to significance are based in Mississippi: *Monsanto Co. v. Scruggs* (hereafter *Scruggs*) and *Monsanto Co. v. McFarling* (hereafter *McFarling*).[5] Whether due to the prevalence of infringers or some other factor, over 10 lawsuits have been filed against farmers in Mississippi (extrapolated from CFS, 2005). In many ways *Scruggs* and *McFarling* are interconnected — even literally, as McFarling purchased seeds from Scruggs Farm Supply in 1999 — and aspects of the cases unfold in reference to each other as they proceed through various levels of court. Unlike *Schmeiser* and *Hoffman*, which tackled the issue of GM contamination (albeit from opposite sides of the liability question), *Scruggs* and *McFarling* rest squarely on the issue of patent infringement from resaving seed originally purchased from Monsanto. Both of these cases appear unambiguous with respect to the evidence of infringing action taken by the farmers; both would appear to be about the process of meting out appropriate punishment for such actions; and yet both provide a determined challenge to the acceptability of classifying seed saving as an infringing action. One significant difference between the cases is that only Homan McFarling signed a TA, allowing legal insight into the legitimacy of the contract itself.

Monsanto Co. v. McFarling

The facts of the charge of saving patented seeds against Homan Mc-Farling are straightforward. McFarling is a soybean farmer in Pontotoc County in northeast Mississippi. In 1997, McFarling bought 200 bags of Monsanto's new RR soybeans. He liked the seeds and thought they saved on labor, so he saved their progeny and planted them in 1998, in conjunction with a further 1,000 bags that he purchased. In both instances, McFarling signed a copy of Monsanto's TA. He saved and cleaned 1,500 bushels of the resulting 1998 crop and used these seeds in part to plant his 1999 crop, and did the same again in 2000. According to McFarling, seed saving has a long tradition in his family:

> We've always saved seed and replanted it. . . . My dad saved them before that, and his dad saved them before that. We've always saved seed. (MS#16, McFarling)

Seed saving was thus part of McFarling's standard agricultural practice, and he claims that when he went to the seed dealer that first year to get a few acres of the new GM seeds for a trial, he wasn't aware that he wasn't supposed to save them. He was presented with a TA at the seed dealer's, but he claims he was unaware of its contents. He was extremely busy, given that it was planting time, and never read the contract: "They said I had to sign it to get the seed, so I just signed it" (ibid.). If this is indeed the case, he wasn't allowed to remain ignorant for long. As early as 1998, Monsanto approached McFarling about his infringing actions, attempted to obtain his records, and proposed to settle with him for approximately $130,000. McFarling refused: "I told them right then, no, I don't got that kind of money. And I didn't want to settle with them; I didn't think I'd done nothing wrong. You know, planting and saving seed, what did I do?" (ibid.). When asked whether it was explained to him what he was considered to have done wrong, he replied in the affirmative, but his position remained unchanged: "Well I still didn't think I was doing nothing wrong. I still don't think I done nothing wrong now" (ibid.). In 2000, Monsanto filed suit against McFarling for patent infringement and breach of contract.

The McFarling case lacks some of the complexity of the Schmeiser case in that McFarling never denied saving the patented seed. It more than makes up for this, however, by raising a wealth of issues with respect to both the patents and the contracts. Further, there were some chronological complexities to the case. This chronology included efforts to change the jurisdiction of proceedings, a stay of proceeding pending a related Supreme Court decision, summary judgment on some claims, and claims held in abeyance while others were forwarded for appeal, some all the way to the Supreme Court.[6] McFarling applied for a U.S. Supreme Court hearing once on the issue of jurisdiction, and again on the issue of patent validity. Both applications were denied and the case went back to the circuit court on the issue of damages. Given these complexities, the following discussion proceeds by issue, allowing for some chronological shifting. These issues span from the extremely abstract (e.g., whether the patent includes the trait and the germplasm or only the trait) to the very concrete (e.g., damages), but *McFarling* consistently questions whether the new system of plant patents and TAs are unfairly biased against farmers. This questioning ultimately draws in the whole agro-biotechnology delivery structure, through patent misuse and antitrust legislation. Notably, the arguments in *McFarling* once again attack the validity of restricting seed saving.

While McFarling did not deny seed saving, he relied on a number of strategies, applications (e.g., regarding venue), and counterclaims (e.g., patent misuse and antitrust) in his defense against the charge of infringement. Given the centrality of the patentability of plants to the issues in *McFarling*, the district court granted a stay pending the Supreme Court *J.E.M. Ag Supply* ruling on the patentability of plants (ultimately upheld in December 2001). Consistent with his values, however, and despite the severity of his legal troubles, McFarling nonetheless indicated to the court that "unless enjoined he intended to plant soybeans saved from the 2000 harvest in 2001."[7] In consequence, Monsanto applied for—and was granted—a preliminary injunction preventing him from doing so. In practice, McFarling's seed saving was clear resistance to the expropriationist rules of the biotech regime. Having refused to settle, however, he moved the issue from one that might be characterized as "training the locals" in the new seed-saving rules to one that directly challenged these rules' legitimacy.

The main claims of the case were finally brought to a hearing on a motion for summary judgment on November 5, 2002. Monsanto applied for this on their claims of patent infringement and breach of contract.[8] While the judge provided no written opinion, the court's reasons can be found in the hearing proceedings. Judge Perry granted summary judgment on the infringement claim, as the "undisputed evidence shows that Mr. McFarling infringed the patent by saving and replanting the seeds."[9] With respect to breach of contract, summary judgment was granted for liability only, with damages remaining to be tried.[10] Judge Perry also granted summary judgment in favor of Monsanto on all of McFarling's defenses and counterclaims, which included claims of violation of the PVP Act, monopolization, unreasonable restraint of trade, and violation of Mississippi antitrust law and patent misuse.

As the PVP Act argument suffered a significant setback with the decision in *J.E.M. Ag Supply*, the weight of McFarling's arguments fell on the issues of antitrust or, at the very least, patent misuse: whereby the "patentee has impermissibly broadened the scope of the patent grant with uncompetitive effect."[11] McFarling argued that the TA created an illegal restraint on trade, in violation of the Sherman Act. Specifically, McFarling argued that its prohibition on seed saving constituted an illegal tying arrangement, as the seed and the trait were really two separate products that operate in separate markets, but a farmer is forced to buy both if he wants one. Rather than purchase the trait anew each year through a royalty payment to Monsanto, and still have the option of saving their own germplasm, farmers must also purchase seeds anew every year. In this way they are forced to buy a product they may not want or need (new seed) in order to get one that they do (a license to use the patented trait). Tying provided a windfall to seed dealers, who gained a captive market.

On the basis of its physical attributes, the tying argument would seem counterintuitive given that the seed and the trait are materially joined. Consistent with this view, Monsanto argued that they are one product and cannot be separated. The issue becomes less obvious when the broader structure of the trait's dissemination is considered, however. Monsanto actually only owns a portion of the seed companies that sell their GM seeds. They additionally license their traits to about 200 seed companies, who put the trait in their seeds and sell them, collecting a royalty on Monsanto's behalf. In fact, in Monsanto's bid to ameliorate monopoly concerns raised by their attempt to purchase a cottonseed company, Monsanto itself argued that the seed and the trait were separate.[12] Consequently, tying is not an issue that can be strictly determined on the technology's physical attributes. Rather, it is an issue that is socially constructed, and important decisions regarding antitrust and patent misuse rest heavily on the nature of this construction.

While arguments on either side become highly abstract, the problem is extremely concrete for farmers. First, there is the additional cost of the seed-saving restriction. If a farmer like McFarling were to save his own soybean seeds, he would only need to pay the approximately $1/bag cost of cleaning them plus the $6.50/acre technology fee. Under the current system, however, a farmer must pay the $6.50 and an additional approximately $25/acre for new seeds. For a farmer like McFarling, who farms 5,000 acres, the cost of seed is approximately $37,000 with seed saving and $157,500 without, a difference of $120,500 a year in input expenses. Obviously, some soybean farmers would save only a portion of their seeds, or even none at all, but for many the difference would be significant. Second, and pertinent to all farmers, the elimination of the "secondary market" of saved seed eliminated a significant source of competition for seed dealers. McFarling argued that this unreasonable restraint on trade resulted in artificially high prices, as farmers became a trapped market. In example, the defense presented evidence that RR soybeans in Argentina, where seed saving was allowed, sold for $12 to $15 per bag, versus $20 to $23 per bag in the United States, where it was not allowed.[13] The practical significance of the argument is clear when taken in the context of the interview comments on skyrocketing seed prices.

While there was little doubt among the farmers interviewed that Monsanto had a monopoly, the courts required a more objective calculation. The distribution structure of GM seeds through licensing traits to seed companies complicated this assessment, however. How much of the market did Monsanto actually control? Did Monsanto's market share only include what it directly owned? Did it include the entire market for RR seeds, most of which was sold through licensing arrangements with other seed dealers? For RR soybeans, for example, the different perspective could assign control at 20% versus 65% or 70%.[14] A related question was how much control Monsanto exerted with respect to its licensees—over the percentage of GM seeds they offered for sale; the combination of traits they offered; and the herbicide choices they provided, among other issues. A further aspect of the market control issue was what impact Monsanto's different technology fees and pricing structures had on different markets, whereby Monsanto's decisions could disadvantage certain regions or countries, as in the earlier Argentina example.

Despite the defense's arguments that these counterclaims raised sufficient issues of material fact to warrant a trial, the judge disagreed. Overall, the district court judge found that Monsanto had not impermissibly broadened the scope of its patent, and he did not find sufficient evidence that it had exercised monopoly power or illegally restrained trade. Despite persuasive arguments that whether the GM trait and the germplasm were indeed separate products was an unresolved question of material fact, and thus not a suitable subject for summary judgment, the judge nonetheless ruled as follows:

> These products are not being illegally tied by any actions of Monsanto. They are inherently tied together as one single product, because the genetic trait is contained in the seed, and for there to be a seed capable of being planted, it has to have been propagated after the insertion of the gene, and so there is simply not two separate products.[15]

Underwriting many of the court's conclusions about antitrust and patent misuse was the argument that McFarling could, if he wanted, purchase non-RR seeds, and therefore was not actually constrained to repurchase seeds every year. This needs to be considered in the context of the perceived competitive disadvantage non-adopters face with respect to time, labor, and conventional seed availability, however. Perhaps even more ominous for those concerned with antitrust and patent misuse was the court's perspective on McFarling's right to bring these questions to hearing. Judge Perry found that McFarling lacked standing on the antitrust issue as he was only "an indirect purchaser" and had "not suffered any antitrust injury as required by the statutes."[16] Such a ruling leaves farmers powerless to initiate an investigation into—and perhaps gain remediation for—the inequities they face in their dealings with Monsanto's monopolistic power. There is virtually no chance of such an antitrust claim arising from the court's prescribed scenario of who had standing—the seed dealers—as they were the main beneficiaries of the disputed method of biotechnology dissemination. Similar to the cause of Saskatchewan's organic farmers, it begs the question of how farmers can effect change in a system that appears so stacked against them.

McFarling appealed the motion to dismiss to the court of appeals for the federal circuit, additionally arguing that the prohibition on seed saving violated the doctrine of patent exhaustion on first sale. In this doctrine, the sale of a patented product authorizes its purchaser to use and sell it, and subsequently the monopoly rights of the patent holder are relinquished. Accordingly, McFarling argued that the seed-saving prohibition was unenforceable as the patent had been exhausted once the seed was sold. Monsanto counter-argued that it was within the scope of their rights to prevent others from making new, patented seeds from their purchase. The court rejected McFarling's argument on the basis that the "first sale" doctrine was "not implicated, as the new seeds grown from the original batch had never been sold,"[17] but had been reproduced on-farm from the first generation of patented seeds. This line of argument raises some interesting (if abstract) questions about the inevitability of its associated conclusion, as the logic would seem to equally support an opposing decision. For example, if the trait in the second-generation GM seed had indeed "never been sold" (and thus had not exhausted the patent placed on the first generation), can it even be said to have been patented at all (again, as the patent was placed on the first generation)?

In another appeals court decision, on April 9, 2004, the court again affirmed the district court's summary judgment on the counterclaims, but not on damages. The defense had reiterated its patent misuse argument that "[by] prohibiting seed-saving, Monsanto has extended its patent on the gene technology to include an unpatented product—the germplasm."[18] The court of appeals acknowledged that the TA's prohibition was actually not restricting the use of goods, but the use of goods "made by, yet not incorporating the licensed good," something which case law had not yet addressed under patent misuse doctrine.[19] Nonetheless, it concluded that given the licensed good and the good made by it were nearly identical copies, "we must presume that Monsanto's '435 patent reads on the first-generation seeds, [therefore] it also reads on the second-generation seeds."[20] The agreement was thus not found to inappropriately extend Monsanto's patent, which read on all generations of the soybeans.

The appeals court did agree that whether the trait and the seed were separate or distinct markets was a question of fact (and thus appropriate for trial, not summary judgment), but it held that this was not relevant to its holding, and consequently "declined to review it."[21] The court also ruled that the antitrust counterclaim failed, because if there was no patent misuse, there was also no violation of the Sherman Act (12). Further, it again dismissed McFarling's arguments concerning the PVP Act in light of the Supreme Court decision in *J.E.M. Ag Supply*, which found that patents and the PVP Act could coexist. In short, McFarling failed to gain ground on any of his arguments, save the question of damages.

Given the newness of GM technologies and the many abstractions involved, there is obviously significant leeway in the application of case law. To date, it appears this leeway has consistently been applied in support of the rights of technology developers and counter to those of farmers. Further, some lack of consistency in the social constructions that underlie decisions—such as the slippage between whether the trait and the germplasm are singular or distinct products—raises equity concerns. Legal abstractions aside, the social decision behind the conflict of rights between technology developers and farmers is clear: does society's need for biotechnological development warrant patents on sexually reproduced plants? If so, does it warrant farmers providing compensation for these patents over successive generations? Even if this were the necessary price for the technologies, the seed-saving restriction remains problematic. Not only does saved seed provide a cap on the "pricing freedom" of seed dealers, but no justification for such post-sale restrictions can be found in patent law (Carstensen, 2006:1072). Moreover, accepting such restrictions has broader repercussions:

> The no-replant policy serves the interest of the patent licensees [the seed dealers] by eliminating saved seed competition, and not the narrowly defined interest of the patent holder. Moreover, the seed companies would have a substantial incentive to standardize on the Monsanto genetic system and not encourage the development of any other systems available. (ibid.)

Carstensen (2006) suggests that if Monsanto can legally prohibit seed saving, it stands to reason that the company could also engage in other restrictions, such as dictating marketing and consequently taxing the sale value of the crop. Essentially, a restriction on seed saving has no legitimate purpose for the monopoly granted to the technology developer, it disproportionately benefits seed dealers at the expense of farmers, and, in the process, it provides avenues for increased market power by the technology developer. Given Monsanto's market share in soybean and cotton crops, this control could have significant repercussions for farmers and farming.

Given the importance and novelty of the issues, those involved in the case considered it to have a good chance for being heard by the Supreme Court, the petition for which was forwarded on the basis of two questions: (1) "may a patent holder lawfully prohibit farmers from saving and replanting seed as a condition to the purchase of patented technology"; and (2) "does obtaining patents on products which are the subject of licensing agreements afford an absolute defense to any claim that the licensing agreements violate the Sherman Act."[22] The Supreme Court invited the acting solicitor general to file a brief expressing the views of the United States on the matter. The acting solicitor general found no reason to disagree with the findings of the appeals court—which was considered to have appropriately rejected the patent misuse and tying claims on the basis of "settled law" — and submitted that the petition for a writ of certiorari should be denied. On June 27, 2005, the petition was denied.

Contracts of Necessity or Crushing Opposition?

Given that McFarling signed the TA and then violated its terms, it would seem that at least this portion of the litigation should be straightforward. It was anything but. In addition to seed-saving restrictions, the contract had two highly controversial provisions: a "forum selection clause," regarding which courts have jurisdiction over disputes, and a "120-multiplier" clause, regarding damages.

The forum selection clause specifies the following:

> The Parties consent to the sole and exclusive jurisdiction and venue of the U.S. District Court for the Eastern District of Missouri, Eastern Division, and the Circuit Court of the County of St. Louis, Missouri, (any lawsuit must be filed in St. Louis, MO) for all claims and disputes arising out of or connected in any way with this agreement and the use of the seed or the Monsanto technologies, except for cotton related claims made by grower. (Monsanto Company, 2006)

Thus a farmer who has a legal dispute with Monsanto, regardless of who initiated it or where the business in dispute was conducted, must travel to Monsanto's hometown of St. Louis, Missouri, for it to be litigated. Such a provision greatly increases the negative impact—not to mention the expense—of any such litigation on the farmer. This clause is no doubt key to 46 out of 90 related cases having been filed in St. Louis. Such a clause is, of course, not binding on those who have not signed a TA. McFarling's defense—that he did not read the contract—is of the utmost practical significance for a busy farmer handed a document while picking up his seed, but is of little legal significance. The question of the fairness of the contract itself increases its significance, however.

Already in 2000, McFarling had attempted to break the clause based on lack of personal jurisdiction of the court. This request was denied.[23] On appeal, in August of 2002, the clause was again considered valid by two of three circuit judges. Circuit judge Clevenger filed a strong dissent, however. He argued that the TA was a "contract of adhesion"—a contract marked by parties with unequal bargaining power, that involves take-it-or-leave-it provisions, that is usually lopsided in favor of the drafter, and where the purchaser has no other source for the necessary goods.[24] Judge Clevenger found the TA to be a contract of adhesion largely because he ascribed greater importance to farmers' use of glyphosate-tolerant seeds. He cited the rapid adoption of RR seeds in only a few years, noted that farmers found the seeds "far more competitive than ordinary seed,"[25] and that Monsanto was the only source for these seeds. "Taken together," he concluded, "these facts indicate that farmers sign the TA if they wish to remain competitive in the soybean market."[26]

In a powerful dissent with the appeals court, Clevenger further outlined in detail how *McFarling* deviated from cited case law, which stipulates that adhesive terms must be "reasonable."[27] While acknowledging the inconvenience to Monsanto of having to litigate against infringers, Judge Clevenger argued that it "has been accepted as another cost attendant to the enjoyment of the patent right."[28] Thus Clevenger asserted that the court had taken an unprecedented step in its decision, which only the Supreme Court should take, and rather bitingly concluded:

> My colleagues have the honor of making this court the first to enforce a forum selection clause in a contract of adhesion against a defendant in derogation of his constitutional rights.[29]

Nonetheless, Clevenger was the minority, and the forum selection clause was upheld.

The liquidated damages provision was another troublesome clause in the TA, which took on a life of its own in *McFarling*. In the 1998 TA, the text of the clause read:

> The grower agrees that damages will include a claim of liquidation damages which would be based on 120 times the applicable technology fee.[30]

Given that Monsanto had no way of knowing whether an infringing farmer would save a limited amount of seed for his own use or would multiply it for an unknown number of users, the company claimed the 120-multiplier to be an approximation of damages. Essentially, Monsanto claimed damages for loss of control over its invention. In the hearing on the motion for summary judgment, Judge Perry supported this logic but ruled against the 120-multiplier, given the legal technicalities of liquidated damages, which were intended to dictate damages that are difficult to measure.[31] The formula proposed by Monsanto required an accounting of actual damages and then multiplied them by 120. These thus constituted punitive damages, which are unenforceable under Missouri law.

Technicalities aside, the clause is of some importance. Monsanto's approach was one that, if legally supported, would presumably terminate the career of any infringing farmer, given that such an award would make bankruptcy virtually inevitable. At the time of McFarling's infringement, Monsanto's technology fee was $6.50 per acre. As McFarling farmed 5,000 acres of land, if he used saved GM seed on all of it without paying Monsanto, the company would lose approximately $30,000 in royalties. A calculation using the 120-multiplier, however, would result in damages of approximately $3.8 million.[32] This would presumably be the end of a farmer like McFarling, who testified that his net worth was approximately $75,000.[33]

Based on her technical concerns with the multiplier, Judge Perry recalculated the liquidated damages and granted judgment against McFarling in the amount of $780,000 for breach of the 1998 contract.[34] On appeal, the circuit court held that the liquidated damages clause was indeed unenforceable, as it did not approximate the amount of harm but was an "invalid penalty clause."[35] The legal basis for this assessment rested on the "anti-one-size-fits-all rule," whereby treating various levels of harm (e.g., saving RR/Bt cotton vs. RR soybeans; or saving for personal use vs. propagating for resale) with similar damages is not a reasonable approximation of damages, but is a penalty. Significantly, this ruling appears to leave open the rewriting of the clause in a more varied—and hence enforceable—manner in the future. According to the court, when such a clause is found unenforceable, only actual damages are available: hence, it vacated the damages and remanded the case back to the district court for a judgment on actual damages.[36] The issue subsequently underwent a district court jury trial, which in June 2005 reached a judgment in the amount of $376,318.00 plus costs of $20,303.25.[37] McFarling appealed the damages (equivalent to $40 per bag, falling between the $6.50/bag royalty and the $70–$80/bag requested by Monsanto),[38] but they were affirmed in May 2007.

The 120-multiplier and the forum selection clause are highly significant to the imbalance of power between biotechnology developers and farmers. Given that Monsanto claims to donate a portion of its litigation proceeds, even it implicitly acknowledges the lawsuits are more about prevention than cost recovery: the 120-multiplier, efforts to triple damages for wilfulness, and the forum selection clause primarily act as an economic deterrent for farmers.[39] Within certain bounds, this is a reasonable means for a business to protect its interests. These bounds appear clearly exceeded, however. While the 120-multiplier was struck down and ultimately removed from the TA, the forum selection clause remains. The ramifications of such imbalanced contractual provisions are increasing as the technology becomes a necessary good. Whatever the legal determination of adhesion, there is no doubt that the TAs significantly disadvantage farmers, who have limited options but to accept them.

Monsanto v. Scruggs

Scruggs had the potential to take many of the issues raised in *McFarling* further in the legal system. Mitchell Scruggs and his brother Eddie Scruggs farmed in northeast Mississippi, with the bulk of their operation in Lee and Pontotoc Counties. Their farm operation was massive, and they were reportedly the largest farmers in three counties, with 20,000 acres under production, 17,000 acres of which were row crop production of soybeans and cotton. The brothers also operated a farm supply business, Scruggs Farm Supply, and a cotton ginning facility. In 2000, Monsanto launched suit against Mitchell and Eddie Scruggs, and their farm supply company (hereafter, "the Scruggs").[40] Monsanto's main claim against the Scruggs was patent infringement related to the alleged use of "brown bagged" (unlicensed) patented cotton and soybean seed containing Monsanto's genetic material in May and June of 2000.

The case had a number of distinctive features. While the Scruggs purchased the GM seeds, they did not sign a TA. Therefore, unlike McFarling, they could not be held to the forum selection clause (consequently, the suit was filed in Mississippi, not Missouri) or the other contract provisions, and the contract's contents was never a subject in the litigation. Second, the fact that the Scruggs had a farm supply outlet cast their activities in a particularly damaging light: if they would violate the seed-saving prohibition, would they also sell the saved seed to their customers? At the very least, the Scruggs are known to have sold Monsanto's GM technologies without making their customers sign the TA, as required by Monsanto. Consequently, there were allegations and counter-allegations regarding the activities that occurred over the years leading up to the lawsuit. Notably, while Monsanto's claim referred specifically to the 2000 crop, the defendants were accused of having saved RR and Bollgard seed in 1997, 1998, 1999, and 2000.[41] Lastly, relative to most other farmers, the Scruggs had access to greater financial resources to expend on litigation. Mitchell

Scruggs alone had an estimated net worth of $5 to $8 million.[42] While far from equivalent to Monsanto's resources, his assets placed him in a significantly better position than most to meet the company in a court of law.

As in *McFarling*, the factual evidence of infringement is fairly straightforward. The Scruggs admitted to purchasing Monsanto's GM soybeans in 1996 and GM cotton in 1998, and saving these seeds for planting in subsequent years through to 2000.[43] In testimony, Mitchell Scruggs disclosed that much of his crop was GM, and that he had purchased initial supplies of GM seeds. The Scruggs vehemently denied that they sold the reconditioned patented seed at their farm supply outlet, however. While *Scruggs* had several unique elements, many of the arguments put forward in it are similar to those in *McFarling*, with a defense once again heavily steeped in the propriety of Monsanto's seed-saving prohibition. In January of 2001, the parties attended a two-day hearing to present witnesses and exhibits in response to the plaintiff's motion for preliminary injunction. Among others, the Scruggs advanced three main arguments in their defense: patent exhaustion, patent misuse and/or antitrust violations, and violation of the PVP Act.

The Scruggs patent exhaustion argument actually predates that in *McFarling*, and is very similar in logic. In *Scruggs*, however, district court judge Pepper did not enter into abstractions over whether the second generation of seed was sold, but took the position that patent exhaustion had no place where the sale was conditional (as the TA made it), something considered valid as long as the conditions are reasonable and not unjustifiably anticompetitive. The judge provided no further treatment of the issue of anticompetitiveness, as he considered the TA to be reasonable. According to Judge Pepper, without the TA, Monsanto's patent would be rendered useless, sales would be reduced, and the broader benefits of technological development might be lost: "[I]t is quite conceivable that it would be commercially infeasible for Monsanto to offer the benefits of its patented biotechnology at all."[44] With respect to the antitrust/patent misuse claims, Judge Pepper concluded that the defendants had not provided sufficient argument for them to warrant his consideration.[45]

Of the defense's three claims, the alleged violation of the PVP Act gained the greatest response from the court. While *J.E.M. Ag Supply* was pending review by the U.S. Supreme Court, Judge Pepper nonetheless had to conclude prior to its outcome. Drawing on the case's earlier circuit court decision, he concluded that the PVP Act was no impediment to enforcing Monsanto's patent. Further, as Monsanto never applied for a PVP Act certificate, it could not be held to its farmers' exemption. This logic is somewhat counterintuitive in its suggestion that not applying for a permit is a viable means of avoiding its negative attributes, and it appears predominantly supported by Judge Pepper's belief in Monsanto's right to utility patents. While he acknowledged there was a deep-rooted tradition of farmers saving seed, he "decline[d] to presume that 'tradition' alone gives the defendants and other modern farmers the unmitigated right to appropriate the plaintiff's technologies to their own use."[46] This declaration of the supremacy of patent holders' rights over those of farmers is unambiguous. It also provided further insight into the judge's motivation for rejecting patent misuse arguments: to allow farmers to refuse Monsanto's seed-saving prohibition would be to allow consumers to "unilaterally determine the circumstances under which a patent holder's discovery is to be offered for use of sale."[47]

Given these conclusions, Judge Pepper granted the plaintiff's motion for a preliminary injunction. Among other things, this injunction prohibited the Scruggs from "all current and future purchase, acquisition, use, sale, offers to sell, transfer (except hereafter provided in paragraph 4), brokering, cleaning, delinting and/or reconditioning"[48] of Monsanto's GM soybean and cotton seeds. The repercussions were obviously going to be significant for the Scruggs' cotton ginning and farm supply business. While acknowledging this impact, the judge ruled that the harm was "largely self-inflicted," and that Monsanto was not obligated to afford the defendants the opportunity to sell its products.[49] As the Scruggs admitted not requiring farmers to sign TAs, and as there were allegations that they sold reconditioned seed, the preliminary injunction was not without cause. The severe effect of such an injunction reflects on both the importance of GM seeds to agricultural production in Mississippi and on the consequent power of its sole supplier, however. Scruggs Farm Supply was ultimately forced to close its doors. In August 2001, the injunction was partially lifted in order to allow the Scruggs to operate their ginning facility and service the needs of local growers. In November 2004, after a number of further summary judgments against the Scruggs, the prohibition against contact with Monsanto's technology was made permanent.

On June 14, 2004, the U.S. district court ruled on a motion for summary judgment on counts 2, 3, 4, and 5 of Monsanto's claims. Evidence was presented that all bags of Monsanto's seed were marked with patent notifications, and that Mitchell Scruggs purchased and cultivated the seed in contravention of this notification: from 1996 to 2000, Scruggs propagated his original purchase of approximately 10 acres of GM soybeans to approximately 8,000 acres worth; and from 1998 to 2000, he propagated a few acres of GM cotton into over 2,000 acres. In response to the "staggering" evidence of infringement, the Scruggs attacked the validity of Monsanto's patents, challenged the scientific rigor of their tests for genetic presence, and pursued their three earlier arguments related to patent exhaustion, violation of the PVP Act, and patent misuse and invalidity. With respect to these, Judge Pepper's opinion was unchanged: he granted in favor of the motion for summary judgment. He did allow that the patent exhaustion doctrine could be affected by antitrust violation, which was a subject for the remaining claims.

These claims were decided just two weeks later. While arguing against the merits of the antitrust claims, Monsanto further argued that the defendants lacked standing, as they had in *McFarling*. This issue was not resolved, however, as the court found the antitrust claims failed on their merits, and rendered the issue moot. With respect to patent misuse and antitrust, the Scruggs' main allegation was that Monsanto engaged in illegal tying arrangements between the GM trait and the seed. Judge Pepper dismissed these allegations as a matter of law based on the *McFarling* federal circuit decision. The Scruggs proceeded with allegations that Monsanto was attempting to implement a seed cartel and to create a market disadvantage for farmers through its web of tying arrangements, incentives, contracts, and other means. They claimed that Monsanto tied the purchase of seed to the purchase of Roundup through its TAs, grower incentive agreements, and seed partner agreements, and further tied the RR and Bt traits in cottonseed.[50]

Many of the arguments resonated with the complaints heard from farmers about the company. The Scruggs' claim was strongest regarding tying Monsanto's RR seeds to its herbicide, given the specific stipulation to that effect in the 1996 to 1998 TAs. This stipulation had been changed by 1999.[51] In 1996 the relevant portion of the agreement read:

> [I]f the Grower uses any glyphosate . . . containing herbicide in connection with the soybean crop produced, from this seed, the herbicide will be a ROUNDUP® BRANDED HERBICIDE (or other Monsanto authorized glyphosate-containing herbicide) labeled for use on ROUNDUP READY (RR) soybeans. No other glyphosate containing herbicide may be used with the patent-protected seed.

In 1997 and 1998, it read:

If a herbicide containing the same active ingredient as Roundup Ultra herbicide (or one with a similar mode of action) is used over the top of Roundup Ready crops, you agree to use only Roundup branded herbicide.[52]

Farmers reading this would obviously conclude that if they use RR technology, they are contractually required to use Roundup instead of a cheaper generic herbicide. Indeed, there was little doubt among the farmers interviewed that this was the case, notwithstanding whether they chose to oblige or not. Monsanto claimed that such restrictions were necessary to ensure compliance with federal regulations that required chemicals be used only according to their labeling, and Roundup was the only chemical labeled for "over-the-top" use. Monsanto further argued that farmers were not compelled to use Roundup, as they were free to not use any herbicide at all "over-the-top" of their RR crops (which would defeat the purpose of purchasing the higher-priced seed), or they could use a nonglyphosate herbicide "over-the-top" of their RR crops (which would kill both crop and weeds).[53] It would thus seem that the practical standard of "reasonableness" required of the TA to preclude the exhaustion of first sale was not applicable to the further requirement that it also not be anticompetitive: Judge Pepper concluded that the defendants failed to prove that "Monsanto forced farmers who wanted to purchase Roundup Ready seeds to purchase Roundup as well."[54] Farmers would likely conclude differently.

In this context it is not surprising that the Scruggs' remaining claims did not fare any better. The judge pursued a similar logic around choice to contend that seed dealers, like farmers, were not forced to use Roundup, despite licenses that "foreclosed seed partners from using glyphosate-based herbicides other than Roundup."[55] He further found no impropriety in Monsanto's program motivating seed dealers to incorporate its traits into a certain percentage of seed sold. He found against claims that farmers were forced to purchase the significantly higher-priced RR/Bt stacked cotton due to an engineered shortage of straight RR cotton. Similarly, he dismissed the Scruggs' contention that Monsanto's grower incentive program—discussed in the previous chapter—unlawfully required the use of Roundup, characterizing it as a voluntary option.

A full assessment of all the issues raised is beyond the scope of this analysis, but the pattern is clear: numerous claims consistent with concerns raised by farmers in the interviews were rejected by the judge over a lack of evidence or other failings. For example, while the "voluntary" aspect of the incentive program is technically supported, interviews with farmers (and even extension agents) suggest that most feel there is no viable economic alternative to accepting its terms. The high cost of GM seeds, and the particular manner of their pricing, makes operating without the incentive program prohibitively expensive and high risk. Farmers' comments regarding the difficulty of obtaining single-trait or conventional seeds similarly suggest some validity to concerns over seed dealer contracts. Despite these seemingly highly significant factual questions regarding whether Monsanto had indeed "implemented a seed cartel"[56] and was attempting to secure rights "beyond its lawful patent monopoly,"[57] the court found them insufficient to proceed to trial and granted summary judgment on all remaining counts.

The Scruggs appealed the court's judgment, arguing that there were sufficient issues of material fact to preclude summary judgment. Specifically, they further pursued their allegations that in the early 1990s "Monsanto developed a business plan to cartelize the soybean, cotton and other seed markets."[58] They argued that Monsanto "misused its patents to impermissibly exclude competitors in trait and herbicide markets, create and police a seed cartel, raise prices, tie/bundle/leverage separate products, fix pricing components, mandate economic waste, harm competition, restrain trade and extract monopoly profits."[59] Significantly, in its appeal brief the defense argued that it would be able to present new antitrust evidence in extensive testimony from experts not presented in *McFarling*.

Given the importance of agriculture to the state, the attorney general of Mississippi, Jim Hood, submitted a brief in support of proceeding to trial. Hood argued that the Scruggs had presented sufficient evidence of Monsanto's market power and "well documented allegations of disturbing exercises of such power."[60] He also found significant resonance in a number of the Scruggs' tying allegations. For example, while the post-1999 TAs no longer required the use of Roundup, there was evidence that Monsanto still enforced the earlier version of the agreement. Hood also found that the judge's rejection of tying in the context where "forcing is absent as long as purchasers have any other option, even one that makes no economic sense"[61] was an extreme interpretation. He further found no reason for the trait and the germplasm to be tied, as even Monsanto conducted these markets differently in different countries. While a number of the attorney general's concerns echoed the Scruggs' arguments regarding farmers' "choice" in the context of the "indispensable nature" of the RR trait, his main concern was the application of summary judgment in the face of sufficient evidence to bring the issues to trial:

> Monsanto's inefficient and costly no-replant policy imposed on Mississippi and other American farmers has continually evaded judicial scrutiny on its merits—or potential lack thereof—as to whether it violates federal antitrust laws. The time is now ripe for such an inquiry.[62]

According to Hood, creating such extreme requirements to withstand summary judgment could significantly impact the state's economy.

Despite the application, the majority of the appeals court nonetheless affirmed the district court judgment on all counts save the technical propriety of its 2004 permanent injunction. The judgment was not unanimous, as circuit judge Dyk dissented on the issue of tying the trait with the use of Roundup, given his reading of Supreme Court rulings. According to Judge Dyk, the effect of the tying arrangement—whether it prevents illegal conduct (such as an unauthorized use of chemical) or provides ancillary benefits—is not relevant to its antitrust status, as there is no "implied antitrust immunity" brought on by federal law.[63] He noted that it is significant that the TA provisions did not require "a government approved herbicide" but a "Roundup branded herbicide,"[64] an important consequence of which was that "competitors are potentially discouraged from seeking regulatory approval or attempting to have the regulation modified or eliminated."[65]

The Scruggs applied for rehearing in September 2006, but the application was denied. They applied for hearing by the Supreme Court, but on April 16, 2007, this was also denied. In 2009, in light of another court decision material to the issue of patent exhaustion, the Scruggs petitioned for the court to reconsider its 2004 denial of their motion for summary judgment. This application was denied. The court did acknowledge that the issue involved a controlling question of law, however, and left an opening for appeal. On May 4, 2009, the U.S. court of appeals denied permission to appeal, but noted that "Scruggs may raise these issues on appeal from the final judgment or injunction."[66] Ultimately, the Scruggs defense ran its course, and in August 2010 the case was returned to a jury for the awarding of damages. On September 21, 2010, Monsanto was awarded $8.9 million in damages, $6.3 million of which was for reselling seeds (Brumfield, 2011). In sum, in spite of considerable effort and a fair amount of support for the importance of the issues to reach hearing, the Scruggs were thwarted in all their attempts at legal headway. Bits of the long litigation carry on. The Scruggs are asking for a reconsideration of the damages judgment or a new trial. On the other side, on the basis of "wilful" infringement, Monsanto is applying for the damages to be tripled (ibid.).

It should be noted that a number of issues relevant to the cases have not been discussed here. For example, patent validity was an important defense in Scruggs. Specifically, Monsanto's GM crops are covered by a number of patents, but important decisions in *McFarling* were based on the only one of these—the '605 patent—that included a claim to the seed. While decisions in *Scruggs* were based on *McFarling*, Monsanto ultimately withdrew the '605 patent from its claims in *McFarling*. This withdrawal not only became the subject of appeals in *Scruggs*, but also spawned arguments by McFarling's defense that the issues in *McFarling* should be relitigated. The details of this and other arguments are beyond the scope of the analysis here, however. What is relevant in both *Scruggs* and *Mc-Farling* is that the changes associated with GM technologies are far from undisputed, and that—despite the clear difficulties—the transition to the new expropriationist paradigm is not occurring without challenge.

If You Do Not Believe: Challenging the Transition

In short, there has not been any significant resistance to genetic engineering in Mississippi on the basis of environmental, health, or any of the usual concerns raised by concerned NGOs. The only significant objections arose around the issue of patenting seeds. Any objection to the technology is primarily manifested in expressions of anger and frustration over the price of seeds and Monsanto's level of control, but few farmers are returning to conventionals. If illicit seed saving is occurring now, none are admitting it. While all are vague on details, for obvious reasons, there is a strong suggestion that in the early years of the technology's introduction, resistance to seed patenting and the prohibition on seed saving was high, and was very practically manifested in farmers ignoring the prohibition.

McFarling states that when Monsanto's GM crops were first introduced, the company was closely monitoring farmers for seed saving:

> Like back in '99 and 2000, they were everywhere. There was just people riding around, you know, checking. . . . They already had an injunction

on me at that time, so they didn't mess with me. A lot of these other boys they messed with, and all of them settled with them probably, except me and Mitchell Scruggs. (MS#16)

The lawyer that represented McFarling had been involved in about nine other cases that were filed. In the early days, his law office had group meetings with affected farmers, the vast majority of whom chose to settle with Monsanto. A paralegal involved commented that in the majority of cases, farmers saved seed because they felt entitled to what they had bought and consequently felt they owned, and out of resistance to what they felt was a disproportionately high cost for the technology. This resonates with comments from those involved in litigation:

It wouldn't be bad if they would only [have] priced their system right. But patented cottonseed costs about $400 a bag for Monsanto's system. A 50-lb. bag! Fifty pounds of cottonseed if you go to the gin with it you are going to get about $2. You take that same $2 worth of seed and treat it and get it cleaned, it is going to cost about $17, so they got a $20 product and they are charging $400. (MS#37, litigant/producer)

In the competitive context of farming, where non-adoption of a new technology spells a significant disadvantage, a number of farmers expressed clear resentment over their narrow options, and this indubitably influenced infringing actions.

There is also some indication that farmers did not immediately comprehend the significance of the agricultural transformation associated with the new biotechnologies. Farmers and seed dealers were rooted in a particular way of doing things and had seen many improvements and new varieties of seeds introduced over the years. The PVP Act had come into force about 25 years earlier, when many of these farmers were already farming, and it had not radically altered their practices. According to anecdotal evidence, violations of the PVP Act were fairly common. Patented GM seeds with a TA were perhaps given the same cursory nod. As one GM cotton farmer not involved in the litigation characterized it, the restrictions on seed saving were known, but the repercussions of violating it weren't:

They knew, [but] it's just like your mamma telling you "no" when you were a kid: there's "no, don't you do it" and then there's "NO, don't you DARE do it." (MS#26)

In addition to this potential lack of comprehension of the significance of the change was the simple fact of the new technology's necessary introduction during farmers' busiest time. Farmers confronted with a TA during the busy planting season might not even recognize the significance of what they are signing. As McFarling testified, he was unaware of the provisions in the TA, and simply responded to the request of the seed dealer to sign the document: "They just handed it to me and said we've got to sign this to get the seed."[67] There have been indications that some seed dealers, responding to farmhands picking up loads of seed for busy farmers, simply signed the farmer's name on the TA. Monsanto itself engaged in such forgeries. In 2004, for example, Illinois farmer Eugene Stratemeyer was sued by Monsanto in what Stratemeyer alleges was a case of entrapment.[68] He turned his counterclaim into a class action lawsuit against the company under the Illinois Consumer Fraud Act when Monsanto agents admitted to forging his and many other signatures on TAs (Schubert, 2001). Monsanto defended on the basis that Stratemeyer lacked standing and failed to allege potential injury (Moeller and Sligh, 2004:45). The forgery of Tennessee farmer Kem Ralph's signature on the TA was also documented in court, although who committed the forgery remains unresolved.

Assuming the best regarding such instances, it would seem that Monsanto had a job of training the locals to the new reality they wished to create around their GM products. Their training has not been gentle, however, and the company's response to noncompliance has been quick and sharp—some might say ruthless. There are many who believe that Monsanto selected a few farmers to prosecute as examples for the others. According to a litigant in Tennessee who wished to remain anonymous:

> Half the people or more in this country were doing what I was doing, and they just picked a few. . . . When it started they just picked a few of us out of our area. (MS#37, litigant/producer)

For those selected, Monsanto appears to spare no effort and expense in their prosecution. One of the Scruggs' counterclaims against Monsanto cites Monsanto's pre-suit surveillance activities to claim invasion of privacy, trespass, and tortuous interference with contract and business relations, among other things. The most extreme example was Monsanto's purchase of an empty lot across from Scruggs Farm Supply, where they set up a trailer from which investigators could put the business under surveillance using video equipment and binoculars.[69] Investigators further followed the Scruggs' family members and employees and asked to search the trucks of customers. While the claims failed on their merits, the basic facts were not disputed.

In addition to such qualitative information, there is quantitative data regarding judgments, where the correlation between the punishment and the crime appears significantly out of balance. It is not surprising, under these circumstances, that the majority of those who find themselves in legal dispute with Monsanto choose settlements over high-stakes litigation. The amount of these settlements seems to vary significantly. According to more than one litigant, farmers who cooperated with Monsanto's efforts to prosecute others received lower settlements. These lower settlements could sometimes be made up (at least in part) with a commitment to purchase a certain amount of GM product over several years. This practice cannot be independently verified, given the nondisclosure agreements, but at least one litigant in Tennessee offered a fairly detailed example of it.

In short, those involved in litigation express an unequivocal opinion that while many farmers saved seed, Monsanto's strategy was to string a few up and scare the rest into compliance. For the selected farmers, the prospects were grim, with Monsanto acting so ruthlessly in its prosecution that even the innocent could not afford to resist settlement. Given that these concerns are raised by those involved in litigation, they have to be taken with that context in mind, of course. The evidence of the imbalance of corporate power and the manner in which it appears to be used, particularly in the United States, does nothing to assuage concerns, however.

For the most part, Monsanto's strategy has been quite effective. One litigant commented that while he knew many others who saved seed, when the lawsuits in his area started that was the end of it:

> And I know people, they got up and they started hauling their seed to the mill instead of cleaning them. (MS#37, litigant/producer)

There was a sentiment, repeatedly uttered, that there was no limit to what Monsanto would try to do to litigants, and talking about the issue would only incur the company's greater wrath. Publicizing the issue is difficult in this context, with litigants even expressing apprehension talking confidentially to an academic: one litigant who promised to meet changed his mind over such apprehensions; another began to have doubts half-way through our conversation. The significance of this culture of fear is great considering the importance of the transformation that agriculture is undergoing and the huge amount of power that Monsanto wields in this transforming system. For some, the prospect of a lifetime ban from the technology, as can be imposed on infringers, is sufficiently threatening. For others, the imbalance of power is sufficient to end disputes quickly. For a limited few, however, this culture only fuels greater resistance.

Tennessee cotton farmer Kem Ralph, for example, was sued by Monsanto in 2000 for seed saving. Ralph became notorious for being the first person to go to jail over GM seed, when he burned disputed seed in contravention of a judge's orders. As a result of his fight, Ralph was fined nearly $3 million in damages for patent infringement.[70] Ralph is a character of strong beliefs, and while his troubles were certainly exacerbated by his conviction that moral certitude trumps law, his case is nonetheless compelling for the vehemence with which he fought the company. While warned by his accountant that I might be a Monsanto spy, for example, Ralph stated he had nothing to risk in talking. Like Schmeiser, he wanted to fight Monsanto in any way he could, and to broadcast his story as widely as possible. Ralph had regular contact with a local paper's editor in an attempt to publicize his concerns, and even had his story covered by an Australian documentary film crew. His perspective is unequivocal:

> Even though I been in prison, I don't care. I feel honored because I'm fighting these people. (MS#38)

While significantly more restrained in approach, McFarling fought Monsanto in his own way. McFarling claimed he rejected the initial settlement offer because he could not afford it. In 2004, after pre-court arbitration, Monsanto made another offer of $1.2 million: half to be paid over the year, and half to be paid in future product purchases over a number of years. Again McFarling rejected it, he claims on the grounds that he didn't have the required initial installment of $300,000. While not privy to specific lawyers' fees and costs of the out-of-state litigation, it seems plausible that McFarling was gambling with far more than that over successive years of litigation. More pointedly, when asked whether he would have settled if Monsanto had asked for a lower amount, McFarling replied: "No, I don't think so. I ain't never thought about settling." His position on the matter is simple: "I don't feel like I've done nothing wrong." As a case in point, even once his legal troubles were already blossoming and he had had a preview of the grief Monsanto intended to cause him, McFarling nonetheless clearly stated his intentions to continue seed saving unless the court ordered him otherwise.

Monsanto did not make a settlement offer to the Scruggs. Some might argue that their apparent resistance is consequently nothing other than a desperate tactic in the face of irrefutable patent infringement evidence. The evidence is stacked against this, however, given the apparent consistency of Mitchell Scruggs' actions and opinions over the course of his interaction with Monsanto's technologies. From his deliberate cultivation and multiplication of Monsanto's patented seeds, to obtaining seed without signing a TA himself, to later refusing to require farmers to sign TAs at Scruggs Farm Supply, Scruggs showed no interest in appeasing Monsanto and every interest in reclaiming a farmer's right to farm saved seed. Pointedly, the Scruggs gained no financial or other benefit by not requiring their seed customers to sign Monsanto's TA. Rather, Mitchell Scruggs acted on his strong feelings about the matter:

> It was no law or anything said that I had to force a farmer to sign a contract that I didn't believe in myself. I mean Monsanto just wanted the farmers to sign it so they would have some kind of strong arm on them. (MS#15)

I interviewed Scruggs in February 2005, soon after the opening of his new store — Scruggs Farm, Lawn and Garden Home Improvement Warehouse — a huge facility that reinvented the Scruggs' business without seed sales. At that time, Scruggs cast his defense broadly, situating his legal issues in the collective concerns of farmers and humanity more broadly:

> I don't think it's fair, then or today or anytime, for one company to use any type of technology to monopolize the whole seed industry and control the food and fiber of the world. That wasn't what patents were intended to do. To be misused and monopolized. (MS#15)

There is little doubt that Scruggs cast himself in a defensive role against these changes. His intention to resist was clear, and it is not surprising that Monsanto went after him with such vehemence.

In sum, while farmers initially resisted the new paradigm prohibiting seed saving, active forms of resistance quickly disappeared in the face of Monsanto's swift reaction. Although strong objections to Monsanto's style of ruling remain, and may one day press forth new resistances, practical manifestations of it currently appear limited to those who faced legal action and refused to settle. For those few, the obvious question is whether their resistance had any broader impact. To a certain extent, it had some effect on publicizing the issues, although certainly not to the extent that occurred in Canada. It also had some effect beyond publicizing the issues, however.

By far the greatest impact of McFarling's counterclaims against Monsanto was the removal of the 120-multiplier. After the circuit court ruled that the clause was unlawful, it was removed from Monsanto's TA. The revised language on remedies is as follows: "patent infringement damages to the full extent authorized by 35 U.S.C. S 271 et seq. . . ." (Monsanto Company, "2006 Monsanto Technology/Stewardship Agreement"). Farmers facing 120 times the damages were under far greater pressure to accept a settlement offer, and the removal of the 120-multiplier clause thus diminished a significant economic threat wielded against alleged infringers. According to a paralegal involved with such litigation:

> A lot of the farmers settled with Monsanto because of the 120-multiplier. . . . Nobody could pay it. That I'm sure made a lot of people settle. (MS#18)

Similarly, the prospect of such legal challenges was likely instrumental in the removal of the contentious glyphosate and trait-tying clause found in earlier versions of the TA. Of course, the greatest potential for significantly reversing the expropriationism evident in association with biotechnologies lay in the antitrust and patent misuse counterclaims.

While McFarling was not successful in his challenges, in many ways he set the stage for the Scruggs case, and the evidence and lines of argument in *Scruggs* certainly did appear to gain in strength. Solicitor Jim Waide noted that the Scruggs had a much better chance of bringing an antitrust challenge as their resources were much greater, whereas McFarling did not even have a patent lawyer (MS#22). Nonetheless, the Scruggs also failed to gain legal ground. The issue may not escape scrutiny indefinitely given (1) the differential structure of technology dissemination in other countries (such as Argentina), where the trait and the germplasm are not tied; (2) the rising concern over the issue by agricultural economists, the broader farming community, and some sectors of government (notably the solicitor general of the United States and the attorney general for the State of Mississippi); and (3) the strong legal challenges seen here. In point of fact, in 2009, the antitrust division of the U.S. Department of

Justice and the USDA announced a series of workshops to explore "the appropriate role for antitrust and regulatory enforcement in the agricultural sector" (United States, *Federal Register*, 2009). Whether such discussions result in substantive change or are merely the antitrust equivalent to green-washing remains to be seen.

Conclusion: The Still Shifting Legal Terrain?

The evidence from Mississippi points to an ongoing reorganization of agriculture, where farmers' property rights are transferred to biotechnology companies, while their ability to control their own farming practices is restricted. The judicial support of utility patents on plants initiated by *Chakrabarty* has ultimately resulted in a proprietary loss for farmers, whose ownership over the progeny of their crops has been expropriated in favor of the full patentability of sexually reproduced plant forms. As the shift to production based on patented GM varieties has been almost wholesale for key agricultural crops, the legal and control mechanisms associated with GM crops, in conjunction with their firm position on the technological treadmill, work together to expropriate Mississippi producers' control over their seed.

Monsanto, specifically, has created a web of contracts and incentives that seemingly dictate many of the options for farmers while creating lucrative opportunities for seed dealers. Without resistance, the further entrenchment of GM technologies is leading to an inevitable shift in control over agriculture from farmers to technology developers. Given the economic imbalance between the two, resistance at the farm level has been quickly curtailed. The punitive provisions of the TA have further hindered resistance. Given the difficulty farmers have in gaining standing for making antitrust arguments, resistance on a broader scale is even more difficult.

As we have seen in *McFarling* and *Scruggs*, however, there is nonetheless some resistance. While its impact has been modest, it has also exposed many socially relevant issues. Notably, *McFarling* and *Scruggs* have raised more antitrust and patent misuse questions than they have answered. Despite its very practical repercussions, for example, the abstract question of whether the trait and the germplasm are separate or unified has produced a delicate (and seemingly contradictory) legal balance. The trait is distinguished from the germplasm in the construction of the patent and in the licensing and dissemination of the technology. Yet a farmer cannot save seed and pay a separate royalty to Monsanto, and claims that this is a tying arrangement are effectively refuted on the grounds that the seed and the trait are inseparable. Given the newness of the issues and the gaps that have to be filled in applicable case law, there is some apparent flexibility in how the many abstractions are handled, and for whose benefit. Resistance in the legal forum has some effect on revealing this process.

In sum, significant issues with respect to farmers' loss of control have been identified. Some losses are the product of social choices about the value of technology development and the need to reward such development. Others are less clear as to their social utility—such as the profits accrued to seed dealers—and appear to have more to do with solidifying a new regime than social benefit. Given that neither *McFarling* nor *Scruggs* managed to escape summary judgment on patent misuse and antitrust issues, the expropriationist trend as yet appears unabated.

CHAPTER 8

Conclusion

We must not allow our options to be foreclosed by ceding to capital the exclusive power to determine how biotechnology is developed and deployed.
KLOPPENBURG, *FIRST THE SEED*

Introduction

Patents were created for the mutual benefit of inventors and society more broadly: through the incentive of monopoly control, they stimulate innovation, research, and development. It is a quid pro quo relationship: at the end of the patent's term, the benefits of the invention are available for the common good. This relationship was conceptualized long before patents on life forms became an issue. Now that they have, the quid pro quo seems to have been questionably translated. If GM life forms can be patented inventions, they fundamentally differ from earlier inventions by being self-reproducible. Turning seeds into patented inventions has consequently facilitated a social reorganization of agricultural production, one of the most essential of human industries.

While the adoption of GM seeds results from the choices of individual farmers, the social reorganization of agricultural production is a result of the proprietary context of these seeds. The paradigm of proprietary agriculture that found its footing in hybrid seeds and the PVP Act in the United States has reached its climax in patented GM seeds. The institutional support for this paradigm has developed more slowly in Canada, but its trajectory clearly follows U.S. trends: through patents, TAs, and the growing body of court decisions, the proprietary emphasis on GM seeds has been increasing in both nations.

The consequences of this emphasis are increasingly evident in, among other things, declining producer control over production decisions and increasing industry-driven crop development. There is strong evidence that the transition to GM technologies is indeed reorganizing the agricultural sector in such a way that farmers will likely only face further reductions in production choices and an increasingly impeded ability to affect the terms and conditions under which they produce. That the technologies can be applied in ways that bolster the already strong linkages between input suppliers and processors only exacerbates the situation. While government regulation and investment into public breeding could offset the negative impacts of this proprietary shift, there has been little evidence of such counterbalancing in either Canada or the United States. Further, this proprietary-oriented reorganization of agricultural production has not resulted from a national inability to regulate in the face of international trade agreements; it is clearly state supported.

What Price Adoption? Expropriationism in a New Era of Agriculture

In both Saskatchewan and Mississippi, the adoption of patented GM technologies has been the result of rational production decisions by individual farmers who assess such factors as whether the technology decreases dockage, increases yields, reduces risk, or provides efficiency benefits. Given the economics of agricultural production, few can afford to make production decisions on any other basis. At the current prices, though to differing degrees depending on crop, a vast majority of farmers are finding that using GM technology provides greater benefits than not using it. Stated somewhat less positively, farmers cannot afford to not adopt any technology that can provide an efficiency gain, regardless of how temporary any associated profit increase may be. In this strictly immediate economic sense, and leaving aside any social, health, or environmental issues, the technology simply offers another tool for the farmer's survival kit.

In Mississippi, many farmers consider the technology their salvation, although this may not last, as insect and weed resistance develop. In Saskatchewan, farmers are slightly less enthusiastic, but still quite keen. Agricultural biotechnologies have provided them with enough of a glimpse of its potential that many hold it out to be their future: whether by drought or frost resistance, or by the creation of new niche markets where they can compete better internationally, biotechnologies suggest a way of staying ahead of the game—or at least in the game. In both Mississippi and Saskatchewan, this game is unequivocally a global one. Biotechnology adoption is coming with a substantial cost, however—one that is over and above the technologies' ticket price.

In the United States, from the first patent on an oil-eating microbe, the legal chronology toward patenting seeds and ending the farmers' exemption with respect to seed saving has been unwavering, and the "anything under the sun" dictum has held sway despite efforts to qualify it. Canadian patent offices and courts have historically been much more reticent about patenting life forms, and they have overseen a progressive qualification of the definition of invention that ultimately excluded higher life forms in *Harvard College*. *Schmeiser* represented a significant departure, however. Before *Schmeiser*, "[m]ore man-made things . . . remain[ed] unpatentable under the Canadian than the American sun" (Vaver, 2004:159); post-*Schmeiser*, the patent regime in Canada is much more in line with that of the United States.

The most significant result of this institutionalization of patent protection on life forms for agriculture is, of course, farmers' loss of their right to save seed. This commodification of the seed locks in high-capital farming with expensive inputs and eliminates an important economic strategy for farmers, particularly in the face of rising seed prices. As these inventions are self-reproducing, numerous questions have arisen from the uncertain legal terrain. While this terrain is still forming, to date the legal decisions in both regions have largely favored the rights of technology developers over those of farmers. In Saskatchewan, liability is increasingly disassociated from ownership, so that the benefits of the technology accrue to its developers while the costs (e.g., contamination, monitoring, and removing volunteers) are borne by its detractors (e.g., organic farmers and non-adopters). In Mississippi, control issues are exacerbated by Monsanto's near monopoly, and case law has only supported Monsanto's control strategies.

The ultimate expression of proprietary control is found in Monsanto's infringement suits against farmers. Given early uncertainty regarding the patentability of life in Canada, the low number of Canadian versus American suits is understandable. Post-*Schmeiser*, they are likely to increase. Infringement lawsuits over self-reproducing technology in themselves require thoughtful consideration. Given farmers' economic and power disadvantage in legal contests with biotechnology companies, these conflicts need clear and balanced rules, with rigorous standards for assessing violators, preferably by a neutral institutional arbitrator. Considering the ruthlessness or comprehensiveness (depending on your perspective) with which Monsanto pursues suspected infringers, discretion in such things as sampling procedures, the percentage of GM presence that is litigable, and the removal of volunteers is highly problematic. Technology developers' need to protect their investments has to be balanced against the rights of farmers. The reports of surveillance, signature forgery, letters to seed dealers, and the like are further troubling in this context of power imbalance, especially given the obligatory nondisclosure agreements for settlements and even for the removal of volunteers.

Expropriationism is evident not only in strengthened intellectual property rights and evolving case law, but also in a whole host of control issues associated with GM technologies. These have been brought to light with the greatest clarity in Mississippi. In Saskatchewan, producers have three herbicide-tolerant canolas to choose from, but in Mississippi, herbicide tolerance means one-stop shopping—at the Monsanto Market. Consequently, as the transition to GM varieties has become the new reality of cotton and soybean farming in Mississippi, farmers have been reduced to dealing with one company, and that company decides rules, contracts, loyalty schemes, and, of course, prices. Monsanto's level of control leaves Mississippi farmers vulnerable to agreements they would likely not accept in a more competitive environment, such as the restrictions on their choice of herbicide (first by TA, and later by "incentive" agreement).

While Monsanto's control is significantly higher in Mississippi than in Saskatchewan, such concentration is intrinsic to the current structure of the agricultural biotechnology sector, and likely to be replicated in new GM applications. In such a context, choices that a technology developer like Monsanto makes can have an enormous impact on farmers. Monsanto's decision to sell the technology and the trait together instead of licensing the technology separately, for example, cost farmers not only their ability to save seed but also the competitive effect this saved seed had on controlling seed prices. Thus Mississippi farmers were greatly disadvantaged by a decision that had no immediate effect on Monsanto's profits, albeit market benefits likely remain to be exposed.

While the control aspects of the technology are much more pronounced in Mississippi than in Saskatchewan, there has been a progressive shift of control in both regions. The evidence from even just these two regions indicates that this use of legal means is not trivial, but is becoming a fundamental component of a new regime of agricultural production. Patents are the latest and most thorough of seed commodification trends, already seen in hybrid technology, plant breeders' rights, and specialty production on contract. The institutional support for the proprietary changes associated with GM technologies provides the ultimate "social" solution to the commodification of the seed, in the manner conceptualized by Kloppenburg (2004). As we have seen in the present study, given the breadth of legal mechanisms affecting control, expropriationism is not just limited to the commodification of the seed but includes systems of power and control associated with the seed to enhance this accumulation. Thus while biotechnologies provide a continuation and a strengthening of the accumulation strategies of appropriationism and substitutionism, they have also introduced new elements into the relationship between capital and agricultural producers.

Some maintain that as long as farmers are voluntarily adopting GM technologies, their control is not really being expropriated, but is freely traded for the benefits they gain. For most farmers, however, the technological treadmill is a reality to which they must comply if they are to survive. Once they've complied, the road behind them disappears with their increasing technological dependence. Consequently, the concept of expropriationism is strongest when biotechnologies' legal aspects are taken in the context of their role in agricultural production. Whether GM crops are, in fact, "better" is not the issue here—and there is significant evidence that short-term benefits are already turning into long-term disadvantages, such as through increasing glyphosate resistance. In the short term, those who do not adopt can suffer a competitive disadvantage with respect to time, labor, and acres they are able to farm. Consequently, GM crop adoption is high enough in the crops investigated here that they represent a clear step on the classic technological treadmill. In a context where biotechnology is becoming the new reality of farming, the impact of a withdrawal of permission to use the technology (for example, for a perceived violation) is itself revealing of the control issues involved.[1] Therefore it is in the interaction between legal and technological, between the prohibition on seed saving and the economically driven adoption of the technology, that expropriationism as a concept is most compelling.

Evolution of the Global Food Regime

Following Gouveia (1997), this research intends to both document difference and identify broader patterns. While there are indeed some differences in the two regions studied here, there are also clearly discernible patterns that make it possible to hypothesize future developments. As farmers transition to more and more GM technologies, viable conventional varieties will become increasingly scarce, and technological dependence on GM varieties will increase. Thus if the proprietary and control issues associated with biotechnologies were to continue rather than face some sort of tempering—for example, through a reinvigoration of the public plant breeding programs—expropriationism will only deepen. A widespread transition of farmers to the role of contract laborers, land managers for mega corporations, or glorified sharecroppers is not far-fetched.

Ultimately, the decline in producer control has repercussions for society more broadly, as a small number of private enterprises dictate who will produce what, and how. Environmental, social, and agricultural benefits can be undervalued or even abandoned in a system where corporate dominance increases the power of such companies to determine what farmers produce and what consumers eat. Nor are expropriationist tendencies limited to agricultural production per se, as it goes hand in hand with larger processes of accumulation by dispossession (Harvey, 2003) evident in bioprospecting and other acts of genetic "enclosures" (Kloppenburg, 2010). This drive to privatization limits access, stifles research, and restricts our potential to deal with future challenges, and consequently, "most of the world's population is rendered the poorer as enclosures of genes and ideas" empower only a "narrow set of decision-makers" to define our available options (ibid.: 384). No better example of the social ramifications of this can be made than the "agrofuels" response to the energy crisis (see, e.g., McMichael, 2009).

While some intellectual property protection on plants is indeed necessary for WTO compliance, the application of this protection to the extent provided by the United States and Canada is not. Despite their interest in providing it, both countries remain reluctant to address the growing number of lawsuits and socially complex legal issues that result. Even the courts have raised questions about the suitability of their forum for addressing biotechnologies' broader social issues. In Canada, for example, both the *Harvard College* and *Schmeiser* Supreme Court decisions acknowledged the existence of unresolved social issues that were the purview of Parliament. Without such input, the courts are restricted to the legalities of the Patent Act.

Is the expropriationism evident here becoming "expropriationism writ large," in a globally replicated, corporate biotechnology backed, neoliberal food regime? Barring some major disruption of the international diffusion of GM technologies—not impossible, given market volatility, liability concerns, and the rush to commercialize socially contentious crops—the concerns raised here have global significance, with an additional emphasis on food security concerns for developing countries. Corporate biotechnology, as a neoliberal globalization "project," would indeed appear to be advancing with some success. While global regulation and integration of food and agriculture production are not inevitable, they are nonetheless making much progress—progress in which biotechnology plays a large part. The third food regime would indeed seem to be constituted by a state-supported international structure of intellectual property rights that facilitates local-level expropriationism and agricultural production based on biotechnological "packages" produced by a highly concentrated corporate sector.

If globalization and the food regime are indeed contested projects, contestation over biotechnologies' role in them is not coming from within the governments of the United States or Canada. Of course, GM technologies do face opposition—for example, through a wide range of protests, acts of GE crop destruction, creation of GE-free zones, various social movement actions, and the legal efforts we have seen here. As noted, the institutionalization of resistance increases the potential for such opposition to ultimately impact the shape of the food regime (Friedmann, 2005; McMichael, 2005). The EU itself provides institutional support for an alternative to many aspects of the U.S.-style biotechnology regime. While the EU GM approval process has been restarted, its new regulatory framework has had many important ripple effects. Labeling legislation, for example, has allowed EU citizens to exert their non-GM preference in the market. In consequence, powerful food processors "are seeking GM-free ingredients and quietly dissuading farmers from growing GM crops" (Kurzer and Cooper, 2007:1055). Thus, local-level agitation has worked itself up the institutional ladder in the EU. This is even further evidenced by the ultimate amendment to Directive 2001/18/EC, which allows member states to restrict or prohibit the cultivation of GMOs in their territories. As Kurzer and Cooper (2007) argue, in the EU "the preferences of citizens have molded the voting behavior of national representatives" (1052).

In a less receptive regulatory environment such as the United States, the greatest potential for orchestrated impact may well lie in more direct access to institutionalization that can occur in the legal forum. In the legal forum there is the potential for judicial decisions counter to the unfolding pro-development biotechnology regulatory paradigm. Through this forum, there is also the potential for broader and longer-term influence. The legal action organized by the SOD is certainly one of the most pro-active acts of legal resistance to the industry. Most recently, we see the suit launched by the Public Patent Foundation in the United States (see Chapter 2), which follows directly on this issue of involuntary contamination. Such offensive action is rare, however, and it is largely in infringement counterclaims that we see the struggle for control over agriculture unfolding.

There have been modest successes: McFarling's blow to the 120-multiplier and Schmeiser's blow to damages based on profit. A difference of just one vote in the 5-4 Supreme Court of Canada *Schmeiser* decision, which would have invalidated plant patents in Canada, would have radically affected Canada's attractiveness for research, development, and production of GM technologies. Similarly, should either McFarling's or the Scruggs' antitrust and patent misuse challenges have found legal purchase, radical changes to the technology's dissemination in the United States would have had to ensue. Of course, there are drawbacks to litigating as a means of social change, as "litigating means that the resolution of a problem is dependent in large measure on the initiative, resources, and tenacity" of both parties (McLeod-Kilmurray, 2007:189). Farmers are out-resourced on all fronts, except perhaps tenacity (for the few individuals in the cases investigated here), and that at no small personal cost.

Such legal acts of resistance can also strengthen rather than weaken the rights of technology developers. They can have spill-over effects that produce both "allied and opposed movements" (Whittier, 2004:532). In fact, there are suggestions that legal mobilization "tends to generate counter-mobilizations of unique scale and success" (McCann, 2004: 519), and can often have negative results for social movements (McCann, 2006). The topic of litigation as a social movement strategy is thus an important area for development, and one that is undertheorized.

Another potential development in institutionalizing resistance to the privatization drive of genetic resources is the more offensive tactic of proactively developing legal alternatives to privatization. For example, biological open-source strategies, modeled on the legal and operational mechanisms of the open-source software movement, are being advocated by some as a means of "repossession" (see Kloppenburg, 2010).

Perhaps most promising of all, from an institutional perspective, is President Obama's overturning of the Bush-era corporate-friendly antitrust environment, which made it difficult to pursue antitrust cases. In January 2010, the U.S. Justice Department began a formal antitrust investigation of Monsanto and its GM soybeans (Kilman, 2010). Among the concerns is Pioneer H-Bred's allegation that "Monsanto is trying to use gene licenses to limit competition" (ibid.). Such concerns are particularly salient given that Monsanto's RR soybeans go off patent in 2014. This ongoing antitrust investigation has now been joined by an investigation by the Securities and Exchange Commission into Monsanto's dealer incentive programs for glyphosate products (Eden, 2011). It seems likely that the claims of the rare individuals like McFarling and Scruggs who decided to fight Monsanto did much to bring such issues to light.

Thus while expropriationist trends are clear in Canada and the United States, and show further indications of global replication, the culmination of these trends is not inevitable. Rather, it appears that while it is difficult, the culmination of upward pressures—particularly those in the institutional environment—can temper or even shift the nature of biotechnological development and dissemination. Given the importance of agriculture to farmers and consumers, this effort seems warranted. Ultimately, we must not forget that these technologies are a tool, and the role society chooses for them will have significant future implications. As Kloppenburg succinctly states it: "To focus too much on the tools rather than on who is using the tools and for what the tools are being used is to misapprehend the problem" (Kloppenburg, 2004:352).

Log of Interviews

Personal Interviews, Saskatchewan, Canada

Interview Number	Interviewee Name or Descriptor	Interview Date
SK1.	Terry Zakreski: lawyer, Schmeiser, Hoffman	March 16, 2005; August 3, 2005
SK2.	Academic, crop development	March 17, 2005
SK3.	Seed dealer/producer, GMO	March 17, 2005
SK4.	Canola Council of Canada	March 18, 2005
SK5.	President, Back to the Farm Research Foundation (organic)	March 20, 2005
SK6.	Arnold Taylor, president, Saskatchewan Organic Directorate; chair, Organic Agricultural Protection Fund	March 20, 2005; August 12, 2005
SK7.	Producers, organic (farm spouses: SK7A and SK7B)	March 20, 2005
SK8.	Producers, organic (farm spouses: SK8A and SK8B)	March 20, 2005
SK9.	Percy Schmeiser, litigant/producer	March 22, 2005; August 3, 2005
SK10.	Hired hand of Schmeiser, producer, conventional	March 22, 2005
SK11.	Producer, GMO	March 23, 2005
SK12.	Producer, organic	March 23, 2005
SK13.	Producer, organic	March 24, 2005
SK14.	Producer, GMO	March 24, 2005
SK15.	Producer/retailer, organic	March 25, 2005
SK16.	Academic, agricultural economist	August 3, 2005
SK17.	Saskatchewan Wheat Pool	August 3, 2005

Interview Number	Interviewee Name or Descriptor	Interview Date
SK18A.	Saskatchewan Association of Rural Municipalities [SARM]	August 4, 2005
SK18B.	SARM representative	August 4, 2005
SK19.	Producer, conventional	August 4, 2005
SK20.	Seed dealer/GM producer	August 4, 2005
SK21.	Saskatchewan Canola Growers Association [SCGA]	August 4, 2005
SK22.	Agricultural consultant/media	August 6, 2005
SK23.	Producer, organic	August 6, 2005
SK24.	Canadian Wheat Board	August 8, 2005
SK25.	Producer, GMO	August 8, 2005
SK26.	Producer, GMO	August 8, 2005
SK27A.	Ag-West Bio, Inc.	August 9, 2005
SK27B.	Ag-West Bio, Inc.	August 9, 2005
SK28.	Saskatchewan Canola Development Commission [SCDC]	August 9, 2005
SK29.	Producer, GM	August 9, 2005
SK30A.	President, Agricultural Producers Association of Saskatchewan [APAS]	August 9, 2005
SK30B.	Regional agriculture organization	August 9, 2005
SK31.	Communication/research director, Organic Agricultural Protection Fund [OAPF], Saskatchewan Organic Directorate [SOD]	August 10, 2005
SK32A.	Saskatchewan Agriculture and Food [SAF]	August 11, 2005
SK32B.	SAF	August 11, 2005
SK33.	Producer, conventional (unaffected crops)	August 11, 2005
SK34.	Vice president, National Farmers' Union [NFU]	August 12, 2005

Other Interviews re Saskatchewan

Interview Number	Interviewee Name or Descriptor	Interview Date
SK35.	Trish Jordan, communications director, Monsanto Canada (telephone interview)	October 17, 2005
K1.	Saskatchewan producer, GMO (personal interview [Kelowna, BC])	November 22, 2004
K2.	Percy Schmeiser, litigant/producer (personal interview [Kelowna, BC])	November 23, 2004

Personal Interviews, Mississippi, United States

Interview Number	Interviewee or Descriptor	Interview Date
MS1.	James Robertson, lawyer: Kem Ralph, Scruggs, and others	May 23, 2005
MS2.	Agricultural expert, marketing	May 23, 2005
MS3.	Producer, GMO	May 24, 2005
MS4.	Producer, GMO	May 24, 2005
MS5.	Farm Bureau	May 24, 2005
MS6.	Producer, GMO	May 25, 2005
MS7.	Agricultural consultant	May 25, 26, 2005
MS8.	Grain elevator manager	May 25, 2005
MS9.	Agricultural expert (hills region)	May 26, 2005
MS10.	Producer, GMO	May 26, 2005
MS11.	Part-time organic market producer and environmentalist	May 26, 2005
MS12.	Producer, conventional	May 26, 2005
MS13.	Producer, GMO	May 27, 2005
MS14.	Producer, GMO	May 27, 2005
MS15.	Mitchell Scruggs, litigant/producer	May 28, 2005
MS16.	Homan McFarling, litigant/producer	May 28, 2005
MS17.	Seed dealer	May 31, 2005
MS18.	Paralegal: McFarling, Scruggs	May 31, 2005
MS19.	Agricultural expert, cotton	June 1, 2005
MS20.	Agricultural expert, soy	June 1, 2005
MS21.	Agricultural expert, corn	June 1, 2005
MS22.	Jim Waide, lawyer: McFarling, Scruggs	June 2, 2005
MS23.	Part-time small organic producer/media	June 2, 2005
MS24.	Producer, GMO	June 3, 2005
MS25.	Agricultural consultant, cotton	June 4, 2005
MS26.	Producer, GMO	June 4, 2005
MS27.	Academic, rural sociology	June 6, 2005
MS28.	Producer, organic, and sustainable agriculture activist	June 6, 2005
MS29A.	Agricultural consultant/producer	June 7, 2005
MS29B.	Agricultural consultant/producer	June 7, 2005
MS30.	Mississippi Department of Agriculture and Commerce [MDAC]	June 8, 2005
MS31.	Producer, GMO	June 8, 2005
MS32.	Editor, local farm news	June 9, 2005
MS33.	Seed dealer	June 9, 2005
MS34.	Producer, GMO	June 9, 2005

Interview Number	*Interviewee Name or Descriptor*	*Interview Date*
MS35.	Agricultural expert, economics	June 9, 2005
MS36A.	Biotech scientist/environmentalist	June 11, 2005
MS36B.	Entymologist/environmentalist	June 11, 2005
MS37.	Litigant/producer	June 12, 2005
MS38.	Kem Ralph, litigant/producer	June 12, 2005
MS39.	Kem Ralph and informed associates (e.g., accountant)	June 12, 2005

Other Interviews re Mississippi

Interview Number	*Interviewee Name or Descriptor*	*Interview Date*
MS40.	Scott Baucum, trait stewardship manager, Monsanto Company (email interview)	February 5, 2007

Notes

Introduction

1. Monsanto's technology agreements are referred to differently in different regions. In order to be consistent with interview responses, each chapter uses the terminology specific to the region discussed.

Chapter 1

1. Thanks to an anonymous reviewer for *Anthropologica* for this wording.

Chapter 2

1. The safeguard clause allows a member state to "provisionally restrict or prohibit the use or sale of an approved GMO if there is 'new or additional information . . . or scientific knowledge' that gives it 'detailed grounds' that the GMO 'constitutes a risk to human health or the environment'" (Article 23 of the EU Directive 2001/18/EC, as cited in Pew, 2005:16).

2. Available online at http://www.legis.nd.gov/assembly/59-2005/bill-text/ FRDJ0200.pdf.

Chapter 3

1. Telephone communication with SAFRR employee, 2006.

2. Producers secure a premium for producing these specialty canolas.

3. The discovery in Canada of a canola that had become resistant to three different herbicides was given a great deal of coverage by environmental organizations (see, e.g., Beyond Pesticides, 2001).

4. The canola check-off is a per-tonne fee deducted at the point of sale. The SCDC check-off deduction is mandatory, but funds are refundable upon request.

5. The Manitoba website notes Monsanto asserts the conditions of its TUA, but that a grower considering keeping a stand of volunteers should contact Monsanto. It was reported to me that Monsanto's concession was to allow producers to keep the stand, with the purchase of the $15/acre TUA. This choice is difficult if the crop is only partially successful (http://www.gov.mb.ca/agriculture/news/wet/volunteercanola.html).

Chapter 5

1. Approximately three months of negotiations ensued after the original request, and it appeared that further time investment would not be fruitful.

2. These included the Canadian Seed Growers Association, the Canadian Seed Trade Association, the Canadian Seed Institute, and the Grain Growers of Canada.

3. *Re Application of Abitibi Co.* (1982), 62 C.P.R. (2d).

4. The following two paragraphs are drawn largely from Roberts (1999) and Atkinson (2005).

5. *Pioneer Hi-Bred v. Canada (Commissioner of Patents), [1989].* 1 S.C.R. 1623.

6. Monopoly control is granted in part on condition of detailed instructions provided so that at the end of the monopoly term someone "skilled in the art" could replicate the invention. The Supreme Court rejected the deposit of seed specimens as complying with this requirement (Roberts, 1999:34).

7. *Harvard College v. Canada* (Commissioner of Patents) [2002]. S.C.R. 45.

8. *Monsanto Canada Inc. v. Schmeiser* (2001) [2001], 12 C.P.R. (4th) 204 (F.C.T.D.), 2001 FCT 256.

9. Ibid. (Trial Proceedings, June 5, 2000 at 278).

10. *Monsanto Canada Inc. v. Schmeiser* (2001) [2001], 12 C.P.R. (4th) 204 (F.C.T.D.), 2001 FCT 256 at para. 29.

11. Ibid. (Trial Proceedings, June 5, 2000 at 870).

12. *Monsanto Canada Inc. v. Schmeiser* (2001) [2001], 12 C.P.R. (4th) 204 (F.C.T.D.), 2001 FCT 256 at para 53.

13. Facts are as accepted by the Supreme Court of Canada: *Monsanto Canada Inc. v. Schmeiser* (2004) [2004], 1 S.C.R. 902 at para. 59–63, 2004 SCC 34.

14. *Monsanto Canada Inc. v. Schmeiser* (2001) [2001], 12 C.P.R. (4th) 204 (F.C.T.D.), 2001 FCT 256 at para. 12.

15. Ibid. at para. 23.

16. Ibid. at para. 80.

17. Ibid. at para. 83.

18. Ibid. at para. 89.

19. Ibid. at para. 114.

20. Ibid. at para. 115.

21. *Monsanto Canada Inc. v. Schmeiser* (2001) [2001], 12 C.P.R. (4th) 204 (F.C.T.D.), 2001 FCT 256 at para. 26.

22. *Monsanto Canada Inc. v. Schmeiser* (2004) [2004], 1 S.C.R. 902 at para. 60, 2004 SCC 34.

23. *Monsanto Canada Inc. v. Schmeiser* (2001) [2001], 12 C.P.R. (4th) 204 (F.C.T.D.), 2001 FCT 256 at para. 119.

24. Ibid. (Trial Proceedings, June 5, 2000, at 1118).

25. Ibid. at 1108.

26. Ibid.

27. *Monsanto Canada Inc. v. Schmeiser* (2001) [2001], 12 C.P.R. (4th) 204 (F.C.T.D.), 2001 FCT 256 at para. 91.

28. Ibid. at para. 92.

29. Ibid. at para. 125.

30. *Monsanto Canada Inc. v. Schmeiser* (2002) [2003] 2 F.C.R. 165 (C.A.) at 172, 2002 FCA 309.

31. Ibid. at 187.

32. Ibid. at 190.

33. Ibid.

34. Ibid. at 191.

35. Ibid. at 193.

36. Ibid. at 194.

37. Ibid.

38. Ibid. at 196.

39. Ibid.

40. Ibid. at 201.

41. Ibid.

42. *Monsanto Canada Inc. v. Schmeiser* (2004) [2004], 1 S.C.R. 902 at 911, 2004 SCC 34.

43. Ibid. at 903.

44. Ibid. at 919.

45. Ibid. at 931–932.

46. Ibid. at 932.

47. Ibid. at 904.

48. Ibid. at 925.

49. Ibid. at 933.

50. Ibid. at 937.

51. Ibid. at 916.

52. Ibid. at 944.

53. Ibid. at 949.

54. Ibid. at 950.

55. Ibid. at 945–946.

56. Ibid. at 906.

57. *Monsanto Canada Inc. v. Schmeiser* (2004) [2004], 1 S.C.R. 902 at 911, 2004 SCC 34 (Factum of the Intervener, Canadian Canola Growers Association [CCGA], December 2003 at para. 20).

58. Ibid. (Notice of Motion, CCGA Proposed Intervener, September 2003 at para. 4(iv)).

59. Ibid. (Factum of the Interveners, Council of Canadians; Action Group

on Erosion, Technology and Concentration; Sierra Club of Canada; National Farmers Union; Research Foundation for Science, Technology, and Ecology; and International Center for Technology Assessment [hereafter "Council of Canadians et al."], at para. 22).

60. Council of Canadians et al. at para. 93.

61. Ibid. at para. 94.

62. *Monsanto Canada Inc. v. Schmeiser* (2004) [2004], 1 S.C.R. 902 at 911, 2004 SCC 34 (Factum of the Intervenor, Attorney General of Ontario, December 16, 2003, at para. 5).

63. Ibid. at para. 4.

64. *Hoffman v. Monsanto Canada Inc.* [2005], 7 W.W.R. 665, 2005 SKQB 225.

65. The issue in *Hoffman* is different from that of Starlink Corn, as Starlink was unapproved for human consumption and was not permitted for release into the human food chain.

66. *Hoffman v. Monsanto Canada Inc.* [2005], 7 W.W.R. 665, 2005 SKQB 225 (Amended, Amended Statement of Claim, January 10, 2002, at para. 26).

67. Ibid. at para. 10.

68. Ibid. (Cross-examination on the Affidavit of Larry Hoffman by Mr. Kuski and Mr. Leurer, Vol. 1, March 23, 2004, at para. 143).

69. Ibid. at para. 439.

70. Ibid. at para. 154–155.

71. *Hoffman v. Monsanto Canada Inc.* [2005], 7 W.W.R. 665, 2005 SKQB 225 (Cross-examination on the Affidavit of Dale Beaudoin by Mr. Kuski and Mr. Leurer, Vol. 1 at para. 176).

72. Ibid. at para. 188.

73. Ibid. at para. 194, 219.

74. *Hoffman v. Monsanto Canada Inc.* [2005], 7 W.W.R. 665, 2005 SKQB 225 (Amended, Amended Statement of Claim, January 10, 2002, at para. 34).

75. s.6 of the *Class Actions Act*, as cited in *Hoffman v. Monsanto Canada Inc.* [2005], 7 W.W.R. 665, 2005 SKQB 225 at para. 25.

76. *Hoffman v. Monsanto Canada Inc.* [2005], 7 W.W.R. 665, 2005 SKQB 225 at para. 67.

77. Ibid.

78. Ibid. at para. 114.

79. Ibid. at para. 122.

80. Ibid. at para. 171.

81. Ibid. at para. 192.

82. EMPA, 2002, s.15(3)(a)(i), as cited in ibid. at para. 161.

83. Ibid. at para. 169.

84. Ibid. at para. 157.

85. Ibid. at para. 158.

86. Ibid. at para. 326.

87. *Hoffman v. Monsanto Canada Inc.* (2006), SKCA (Application for Leave to Intervene), [2006 SKCA 132].

88. All quotations are taken from letters posted in the archives of Percy Schmeiser's website (available at http://www.percyschmeiser.com/harassment.htm).

89. *Monsanto Canada Inc. v. Schmeiser* (2001) [2001], 12 C.P.R. (4th) 204 (F.C.T.D.), 2001 FCT 256 (Trial Proceedings, 2000 at 1123).

90. Ibid. at 1125.

91. Ibid. at 1109.

92. Ibid. at 1116.

93. Ibid. at 1126.

94. *Schmeiser v. Monsanto Canada Inc.* (2001), SK Small Claims Court, Humboldt, SK, at para. 22.

95. *Monsanto Canada Inc. v. Schmeiser* (2004), 1 S.C.R. 902 at 914, 2004 SCC 34.

96. *Schmeiser v. Monsanto Canada Inc.* (2001), SK Small Claims Court, Humboldt, SK, at para. 42.

97. Ibid. at para. 49.

98. Ibid. at para. 50.

99. *Hoffman v. Monsanto Canada Inc.* [2005], 7 W.W.R. 665, 2005 SKQB 225 at para. 337.

100. Ibid. at para. 335.

Chapter 6

1. See, for example, Neuman and Pollack (2010) regarding farmers' growing battle with glyphosate-resistant weeds.

2. In the United States, the Technology Agreement is officially called the Technology/Stewardship Agreement, but everyone calls it the "Technology Agreement," and that is the convention followed here.

Chapter 7

1. *Diamond v. Chakrabary*, 447 U.S. 303 (1980).

2. *Ex Parte Hibberd*, 227 U.S.P.Q. 443 (1985).

3. *J.E.M. Ag Supply, Inc. V. Pioneer Hi-Bred Int'l, Inc.*, 534 U.S. 124 (2001).

4. It should be noted that Monsanto "failed to participate in the arbitration hearing without good cause." Judge Hoberg, from the State of North Dakota Seed Arbitration Board nonetheless concluded: "The greater weight of the evidence show the Nelson Farm does not owe Monsanto any damage for patent infringement, conversion, or unjust enrichment because the evidence presented shows that none of these occurred. . . . Nelson Farm has presented substantial evidence disproving Monsanto's case." Available online at NelsonFarm.Net, "Recommended Order."

5. *Monsanto Co. v. Scruggs*, 249 F. Supp. 2d 746 (N.D. Miss. 2001); *Monsanto Co. v. McFarling*, 302 F. 3d 1291 (Fed. Cir. 2002).

6. See notes 8 and 10, for example.

7. *Monsanto Co. v. McFarling*, 302 F. 3d 1291 (Fed. Cir. 2002) at 1294.

8. The claims were made for their '435 and '605 patent, but the case proceeded only on the '605 patent after Monsanto withdrew the former. This withdrawal be-

came the subject of arguments regarding the defendant's right to relitigate, given that the '605 patent was the basis of earlier decisions.

9. *Monsanto Co. v. McFarling* (E.D. Mo., April 13, 2000) (Lexis 22259) (Hearing on Motion for Summary Judgment, November 5, 2002:50).

10. Questions were raised around the 1997 contract with respect to the location of the restrictive contractual provisions (McFarling's signature appeared on the front pages; the back pages of the contract were missing). The district court consequently granted summary judgment on claim 3 (breach of contract) with respect to the 1998 contract; the 1997 contract was held while that decision could be appealed.

11. *Monsanto Co. v. McFarling*, 363 F. 3d. 1336 (Fed. Cir. 2004) at 1341.

12. *Monsanto Co. v. McFarling* (E.D. Mo., Apr. 13, 2000) (Lexis 22259) (Hearing on Motion for Summary Judgment, November 5, 2002:20).

13. *Monsanto Co. v. McFarling*, 363 F. 3d. 1336 (Fed. Cir. 2004) (Corrected Brief of Appellant, March 2003:22).

14. *Monsanto Co. v. McFarling* (E.D. Mo., Apr. 13, 2000) (Lexis 22259) (Hearing on Motion for Summary Judgment, November 5, 2002:29).

15. Ibid. at 49.

16. Ibid. at 46.

17. *Monsanto Co. v. McFarling*, 302 F. 3d 1291 (Fed. Cir. 2002) at 1299.

18. McFarling, H. as cited in *Monsanto Co. v. McFarling*, 363 F. 3d 1336 (Fed. Cir. 2004) at 1341.

19. Ibid. at 1343.

20. Ibid.

21. Ibid. at 1344.

22. *Monsanto Co. v. McFarling*. 363 F. 3d. 1336 (Fed. Cir. 2004) (Petition for a Writ of Certiorari, Waide and Associates, 2004 at i).

23. *Monsanto Co. v. McFarling* (E.D. Mo., Apr. 13, 2000) (Lexis 22259).

24. *Monsanto Co. v. McFarling*, 302 F. 3d 1291 (Fed. Cir. 2002) at 1300.

25. Ibid. at 1301.

26. Ibid.

27. The case law basis of the enforceability of forum selection rested on a decision involving Carnival Cruise Lines, which traveled internationally and thus was vulnerable to suits launched in multiple jurisdictions. Clevenger found that Monsanto "can lay claim to few or none of the policy rationales" behind the case. In the Carnival case, forum selection was imposed to create order and certainty, with the savings in litigation costs passed on to passengers in lower ticket prices. Cruise lines also rarely sue their passengers, and if customers object to the contract they can choose another cruise line. In contrast, Monsanto's TA was drafted by the prospective plaintiff (not the defendant), who can choose which cases to pursue. All preceding business was conducted in the locale where the farmer resides and reasonably expects any legal proceedings to occur, and patent law is consistent across the United States. Further, any litigation cost savings are unlikely to be passed onto farmers, as GM seed prices are driven by demand, not production. Finally, farmers cannot obtain the product in question—given Monsanto's virtual monopoly—from any other source.

28. Ibid. at 1304.

29. Ibid. Note 4 at 1306.

30. *Monsanto Co. v. McFarling* (E.D. Mo., April 13, 2000) (Evidentiary Hearing, April 6, 2000:21).

31. Ibid. (Hearing on Motion for Summary Judgment, November 5, 2002).

32. Ibid. at 25.

33. *Monsanto Co. v. McFarling* (E.D. Mo., April 13, 2000) (Lexis 22259) (Evidentiary Hearing, April 6, 2000:12).

34. *Monsanto Co. v. McFarling* (E.D. Mo., November 15, 2002) (Lexis 27289).

35. *Monsanto Co. v. McFarling*, 363 F. 3d. 1336 (Fed. Cir. 2004) at p. 29 (original unpublished document).

36. Ibid.

37. *Monsanto Co. v. McFarling*, 363 F. 3d. 1336 (Fed. Cir. 2004) (Corrected Brief of Appellant, January 4, 2006, at 16).

38. *Monsanto Co. v. McFarling*, 488 F. 3d 973 (Fed. Cir. 2007) at 977.

39. While it is commonly asserted in the farm community that Monsanto donates a significant portion of its legal awards to farm organizations such as the Mississippi Farm Bureau, the company itself only acknowledges that it returns "all pre-trial settlement dollars to the agricultural community" (Scott Baucum, 2007, personal communication, Monsanto Company trait stewardship manager).

40. The named defendants in the original complaint included Mitchell and Eddie Scruggs and Scruggs Farm Supply. By 2002, the defendants in the third amended complaint included Mitchell and Eddie Scruggs, Scruggs Farm and Supplies, LLC, Scruggs Farm Joint Venture, MES Farms, Inc., HES Farms, Inc., and MHS Farms, Inc. In all cases the defendants will be referred to as "the Scruggs."

41. *Monsanto Co. v. Scruggs*, 249 F. Supp. 2d. 746 (N.D. Miss. 2001) (Third Amended Complaint, 2002).

42. Ibid. at 760.

43. *Monsanto Co. v. Scruggs*, 459 F. 3d. 1328 (Fed. Cir. 2006) (Brief of Appellants, May 2, 2005:4).

44. *Monsanto Co. v. Scruggs*, 249 F. Supp. 2d 746 (N.D. Miss. 2001). Note 4 at 753-754.

45. Ibid. at 755.

46. *Monsanto Co. v. Scruggs*, 249 F. Supp. 2d. 746 (N.D. Miss. 2001) at 757.

47. Ibid.

48. Ibid. at 761.

49. Ibid. at 760.

50. *Monsanto Co. v. Scruggs*, 342. F. Supp. 2d. 568 (N.D. Miss. 2004) at 576.

51. By 1999, Monsanto had changed the requirement to include Roundup "or other authorized non-selective herbicide which could not be used in the absence of the Roundup Ready gene" (1999 Technology Use Guide, cited in *Monsanto Co. v. Scruggs*, 342. F. Supp. 2d. 568 (N.D. Miss. 2004) (Lexis 26691) at note 3.

52. *Monsanto Co. v. Scruggs*, 342. F. Supp. 2d. 568 (N.D. Miss. 2004) at 576.

53. Ibid.

54. Ibid. at 577.

55. Ibid. at 578.

56. Ibid. at 580.

57. Ibid. at 584.

58. *Monsanto Co. v. Scruggs*, 459 F. 3d. 1328 (Fed. Cir. 2006) (Brief of Appellants, May 2, 2005:4).

59. Ibid. at 8.

60. *Monsanto Co. v. Scruggs*, 459 F. 3d. 1328 (Fed. Cir. 2006) (Brief of Amicus Curiae, Jim Hood, Attorney General State of Mississippi, May 20, 2005:20).

61. Ibid. at 10.

62. Ibid. at 16.

63. *Monsanto Co. v. Scruggs*, 117 Fed. Appx. 729 (Fed. Cir. 2004) (Lexis 27519) at 1343.

64. Ibid. at 1344.

65. Ibid. at 1343.

66. *Monsanto Co. v. McFarling*, No. 900, 2009, Lexis 11700 (Fed. Cir. May 4, 2009) at 5.

67. *Monsanto Co. v. McFarling* (E.D. Mo., Apr. 13, 2000) (Lexis 22259) (Evidentiary Hearing, April 6, 2000: 19, 16).

68. Stratemeyer alleges that a man approached him when it was too late in the season to start a crop and requested that Stratemeyer sell him some seeds for erosion control. He reluctantly helped him, charging him only enough to cover the cost of cleaning and bagging the seed (Schubert, 2001).

69. *Monsanto Co. v. Scruggs*, 342 F. Supp. 2d. 602 (N.D. Miss 2004) at 606.

70. *Monsanto Co. v. Kem L. Ralph*, 382 F. 3d 1374 (Fed. Cir. 2004).

Conclusion

1. The Nelson Farm case in North Dakota (see Chapter 7, note 4) is illustrative of this. There was ultimately no finding of the Nelsons' wrongdoing. Nonetheless, in the course of their dispute, Monsanto sent approximately 290 letters to seed dealers notifying them that Monsanto did not authorize anyone associated with Nelson Farm to "possess, make, use or transfer" any of their technology. Ultimately this resulted in the termination of a contract to produce 600 acres of RR soybeans for Mycogen Seeds. (Letters available on NelsonFarm.Net.)

Bibliography

General Sources

Anderson, P. 2001. "The GE Debate: What Is at Risk When Risk Is Defined for Us?" *Capitalism, Nature, Socialism* 12(1): 39–44.

Andree, P. 2002. "The Biopolitics of Genetically Modified Organisms in Canada." *Journal of Canadian Studies* 37(3): 162–191.

Aoki, K. 2008. *Seed Wars: Controversies and Cases on Plant Genetic Resources and Intellectual Property*. Durham, NC: Carolina Academic Press.

Arends-Kuenning, M., and F. Makundi. 2000. "Agricultural Biotechnology for Developing Countries." *American Behavioral Scientist* 44(3): 318–350.

Atkinson, R. 2005. "Mixed Messages: Canada's Stance on Patentable Subject Matter in Biotechnology." *Intellectual Property Journal* 19: 1–27.

Barboza, D. 2003. "Monsanto Sues Dairy in Maine over Label's Remarks on Hormones." *New York Times*. Available online: http://www.nytimes.com/2003/07/12/business/monsanto-sues-dairy-in-maine-over-label-s-remarks-on-hormones.html. Accessed August 11, 2011.

Barham, E. 1997. "Social Movements for Sustainable Agriculture in France: A Polanyian Perspective." *Society and Natural Resources* 10(3): 239–249.

Barton, J. H., and P. Berger. 2001. "Patenting Agriculture." *Issues in Science and Technology* 17(4): 43–50.

Bauer, M., and G. Gaskell (Eds.). 2002. *Biotechnology: The Making of a Global Controversy*. Cambridge: Cambridge University Press.

Beck, U. 1992. *Risk Society: Towards a New Modernity*. M. Ritter, trans. London: Sage Publications.

Beingessner, P. 2004. "New Varieties Only Available to Farmers if They Pay Each Year." *Crop Choice News*. Available online: http://www.cropchoice.com/leadstrygmo122704.html. Accessed February 24, 2005.

Berlan, J-P. 1991. "The Historical Roots of the Present Agricultural Crisis." *Towards a New Political Economy of Agriculture*. W. H. Friedland, L. Busch, F. H. Buttel, and A. P. Rudy, eds. Boulder: Westview Press. 115–136.

Bernauer, T. 2005. "Causes and Consequences of International Trade Conflict over Agricultural Biotechnology." *International Journal of Biotechnology* 7 (1/2/3): 7–28.

Beyond Pesticides. 2001. "Genetically Contaminated Superweeds Invade Canada." Available online: http://www.beyondpesticides.org/news/daily_news_archive/2001/02_08_01.htm. Accessed April 16, 2007.

Bjorkquist, S. (under supervision, Winfield, M.) 1999. *The Regulation of Agricultural Biotechnology in Canada*. The Canadian Institute for Environmental Law and Policy. Available online: http://cielap.org/pdf/regbiotch.pdf. Accessed November 11, 2008.

Bourdieu, P. 1987. "The Force of Law: Towards a Sociology of the Juridical Field." *Hastings Law Journal* 38: 805–853.

Boyens, I. 1999. *Unnatural Harvest: How Corporate Science Is Secretly Altering Our Food*. Toronto: Doubleday Canada Ltd.

———. 2001. *Unnatural Harvest: How Genetic Engineering Is Altering Our Food*. Toronto: Doubleday Canada Ltd.

Bratspies, R. 2002. "The Illusion of Care: Regulation, Uncertainty, and Genetically Modified Food Crops." *N.Y.U. Environmental Law Journal* 10: 297–319.

———. 2003. "Myths of Voluntary Compliance: Lessons from the StarLink Corn Fiasco." *William and Mary Environmental Law and Policy Review* 27: 593–649.

Brumfield, Patsy R. 2011. "Monsanto Wants Triple Damages against Scruggses." *Northeast Mississippi Daily Journal* (Tupelo). Available online: http://nems360.com/pages/full_story/push?article-Monsanto+wants+triple+damages+against+Scruggses%20&id=14293306&instance=secondary_stories_left_column. Accessed June 29, 2011.

Busch, L., and C. Bain. 2004. "New! Improved? The Transformation of the Global Agrifood System." *Rural Sociology* 69(3): 322–346.

Busch, L., W. Lacy, J. Burkhardt, and L. Lacy. 1991. *Plants, Power and Profit: Social, Economic and Ethical Consequences of the New Biotechnologies*. Cambridge: Basil Blackwell.

Buttel, F. 1995. "Biotechnology: An epoch-making technology?" *The Biotechnology Revolution?* M. Fransman, G. Junne, and A. Roobeek, eds. Oxford: Blackwell.

———. 2003. "The Global Politics of GEOs: The Achilles' Heel of the Globalization Regime?" *Engineering Trouble: Biotechnology and Its Discontents*. A. Schurman and D. Kelso, eds. Berkeley: University of California Press. 152–173.

Buttel, F., and P. LaRamee. 1991. "The 'Disappearing Middle': A Sociological Perspective." *Towards a New Political Economy of Agriculture*. W. L. Friedland, L. Busch, F. Buttel, and A. Rudy, eds. Boulder: Westview Press. 151–169.

Buttel, F., O. Larson, and G. Gillespie Jr. 1990. "The New Political Economy of Agriculture: An Evaluation." *The Sociology of Agriculture*. Westport, CT: Greenwood Press, Inc. 171–186.

Canadian Biotechnology Action Network. 2011. "U.S. Court Defeats Monsanto's Genetically Modified Alfalfa, for Now." CBAN. Available online: http://www.cban.ca/Press/Press-Releases/U.S.-Court-Defeats-Monsanto-s-Genetically-Modified-Alfalfa-For-Now. Accessed August 20, 2011.

Canadian Encyclopedia. "Agriculture and Food." Available online: http://thecanadianencyclopedia.com. Accessed July 24, 2006.

Canadian Wheat Board. 2005. "Biotechnology Position Statement." Available online: http://www.cwb.ca/en/topics/biotechnology/index.jsp. Accessed June 2005.

Carstensen, P. 2006. "Post-Sale Restraints via Patent Licensing: A "Seedcentric" Perspective." *Fordham Intellectual Property, Media and Entertainment Law Journal* 16: 1053–1091.

Castells, M. 2000. *The Information Age: Economy, Society, and Culture*. 3 vols. 2nd ed. Oxford and Malden, MA: Blackwell.

CBC Television. 2003. "Canadian Government in Conflict of Interest over Genetically Modified Wheat Partnership with Monsanto." CBC Television, *The National*. Alison Smith (host) and David Common (reporter). November 28, 2003. 22:00 EST. Available online: http://www.healthcoalition.ca/gewheat-cbc.pdf. Accessed January 3, 2005.

Center for Food Safety. 2005. *Monsanto vs. U.S. Farmers*. Washington, DC: Center for Food Safety.

———. 2007. *Monsanto vs. U.S. Farmers: November 2007 Update*. Available online: http://www.centerforfoodsafety.org/pubs/Monsanto%20November%20 2007%20update.pdf. Accessed: January 10, 2009.

Clark, E. A. Undated. "The Implications of the Schmeiser Decision." Monsanto vs. Schmeiser Website. Available online: http://www.percyschmeiser.com/.

Clark, G. 1992. "'Real' Regulation: The Administrative State." *Environment and Planning A* 24: 615–627.

Cochrane, W. W. 1979. *The Development of American Agriculture: A Historical Analysis*. Minneapolis: University of Minnesota Press.

Coglianese, C. 2001. "Social Movements, Law, and Society: The Institutionalization of the Environmental Movement." *University of Pennsylvania Law Review* 150: 85–118.

Constance, D., A. Bonanno, C. Cates, D. L. Argo, and M. Harris. 2003. "Resisting Integration in the Global Agro-Food System: Corporate Chickens and Community Controversy in Texas." *Globalization, Localization and Sustainable Livelihoods*. R. Almas and G. Lawrence, eds. London: Ashgate. 103–118.

"Consumers Must Know Value of GMOs." *Star Phoenix*. June 22, 2001. Available online. http://www.monsanto.co.uk/news/ukshowlib.phtml?uid=5220. Accessed April 21, 2007.

Davidson College Website. 2004. "EU vs US: WTO Court Case." Available online: http://www.bio.davidson.edu/people/kabernd/seminar/2004/GMevents/ JA/JAWTO.html. Accessed January 10, 2006.

Desmarais, A. A. 2007. *La Via Campesina: Globalization and the Power of Peasants*. Halifax: Fernwood Press.

Dhar, T., and J. Foltz. 2007. "The Impact of Intellectual Property Rights in the Plant/Seed Industry." *Seeds of Change*. J. Kesan, ed. Oxon, UK: CABI Press. 1–21.

Dutfield, G. 2003. *Intellectual Property Rights and the Life Science Industries: A Twentieth Century History*. Bodmin, Hampshire: Ashgate.

Earl, J. 2004. "The Cultural Consequences of Social Movements." *The Blackwell Companion of Social Movements*. D. Snow, S. A. Soule, and H. Kriesi, eds. Malden, MA: Blackwell. 508–530.

Eden, S. 2011. "Monsanto's SEC Probe Adds to Scrutiny." *The Street*. Available online: http://www.minyanville.com/businessmarkets/articles/thestreet-Mon santo-genetically-modified-seeds-Roundup/6/29/2011/id/35461. Accessed August 23, 2011.

Elias, P. 2006. "Biotech Firm Raises Furor with Rice Plan." Associated Press. May 14, 2006. Available online: http://news.yahoo.com/s/ap/20060514/ap_on_sc/ biopharming_dilemma. Accessed May 23, 2006.

———. 2007. "Biotech Seeks to Ease Fuel's Reliance on Oil, Corn." *The Associated Press*. April 15, 2007. Available online: http://www.journalstar.com/ar ticles/2007/04/15/news/nebraska/doc4621494a6e66b550598981.txt. Accessed April 21, 2007.

Environmental Commons. 2006. "2006 Food Control Legislation Tracker." Available online: http://www.environmentalcommons.org/gmo-tracker.html. Accessed January 18, 2006.

Environmental News Service. 2006. "U.S. Seeks to Remove Biotech Food Labeling from Codex Agenda." ENS Website: http://www.ens-newswire.com. Accessed January 14, 2006.

ETC Group [Action Group on Erosion, Technology and Concentration]. 2003. "Mulch Ado about Nothing? . . . Or the "Sand Witch?" ETC Group Communiqué. September/October 2003. Issue 81. Available online: http://www .etcgroup.org/article.asp?newsid=413. Accessed January 3, 2006.

———. 2005a. "Global Seed Industry Concentration-2005." ETC Group Communiqué. September/October 2005 Issue 90. Available online: http://www .mindfully.org/Farm/2005/Global-Seed-Industry6sep05.htm. Accessed February 11, 2006.

———. 2005b. "Oligopoly, Inc. 2005: Concentration in Corporate Power." ETC Group Communiqué. December 2005. Issue 91. Available online: http://www .etcgroup.org/en/materials/publications.html?pub_id=42. Accessed February 11, 2006.

———. 2008. "Who Owns Nature? Corporate Power and the Final Frontier in the Commodification of Life." Available online: http://www.etcgroup.org/up load/publication/707/01/etc_won_report_final_color.pdf. Accessed August 20, 2011.

Evenson, D. D. 2000. "Patent and Other Private Legal Rights for Biotechnology Inventions (Intellectual Property Rights IPR)." *Agriculture and Intellectual Property Rights: Economic, Institutional and Implementation Issues in Biotechnology*. Wallingford, Oxfordshire, UK: CABI Publishing. 11–24.

Falcon, W. P., and C. Fowler. 2002. "Carving up the Commons—Emergence of a New International Regime for Germplasm Development and Transfer." *Food Policy* 27: 197–222.

Fitting, E. 2008. "Importing Corn, Exporting Labour: The Neoliberal Corn Regime, GMOs, and the Erosion of Mexican Biodiversity." *Food for the Few: Neoliberal Globalism and Biotechnology in Latin America*. G. Otero, ed. Austin: University of Texas Press. 135–158.

Friedland, W. 1991. "Introduction: Shaping the New Political Economy of Advanced Capitalist Agriculture." *Towards a New Political Economy of Agriculture*. W. Friedland, L. Busch, F. Buttel, and A. Rudy, eds. Westview Special Studies in Agriculture Science and Policy. Boulder: Westview Press.

———. 2002. "Agriculture and Rurality: Beginning the 'Final Separation'?" *Rural Sociology* 67(3): 350–371.

———. 2004. "Agrifood Globalization and Commodity Systems." *International Journal of Sociology of Agriculture and Food* 12: 5–16.

Friedmann, H. 1991. "Changes in the International Division of Labour: Agrifood Complexes and Export Agriculture." *Towards a New Political Economy of Agriculture*. W. Friedland, L. Busch, F. Buttel, and A. Rudy, eds. Westview Special Studies in Agriculture Science and Policy. Boulder: Westview Press. 65–93.

———. 1992. "Distance and Durability: Shaky Foundations of the World Food Economy." *Third World Quarterly* 13(2): 371–383.

———. 1993. "The Political Economy of Food: A Global Crisis." *New Left Review*, 197: 29–57.

———. 1995a. "Food Politics: New Dangers, New Possibilities." *Food and Agrarian Orders in the World-Economy*. P. McMichael, ed. Westport, CT: Greenwood Press. 15–34.

———. 1995b. "The International Political Economy of Food: A Global Crisis." *International Journal of Health Services* 25(3): 511–538.

———. 2000. "What on Earth Is the Modern World-System? Foodgetting and Territory in the Modern Era and Beyond." *Journal of World-Systems Research* 6(2): 480–515.

———. 2004. "Feeding the Empire: The Pathologies of Globalized Agriculture." *Socialist Register 2005: The Empire Reloaded*. London: Merlin Press. 124–143.

———. 2005. "From Colonialism to Green Capitalism: Social Movements and Emergence of Food Regimes." *Research in Rural Sociology and Development* 11: 227–264.

Friedmann, H., and P. McMichael. 1989. "Agriculture and the State System: The Rise and Decline of National Agricultures, 1870 to the Present." *Sociologia Ruralis* 29(2): 93–117.

Galanter, M. 1983. The Radiating Effects of Courts. *Empirical Theories of Courts*. Keith D. Boyum and Lynn Mather, eds. New York: Longman. 117–142.

———. 1974. "Why the 'Haves' Come Out Ahead: Speculation on the Limits of Legal Change." *Law and Society Review* 9: 95–160.

Gene Watch UK. "European Food and Feed Safety Regulations." Available online: http://www.genewatch.org/article.shtml?als%5Bcid%5D=405263&als%5Bitemid%5D=507957. Accessed April 21, 2007.

Gillam, C. 2007. "Monsanto Profit Beats Estimates, Raises Outlook." *National Post*. April 4, 2007. Available online: http://www.canada.com/nationalpost/financialpost/story.html?id=93885d91-d5fe-46a2-9fa5-4d9555fd53c8&k=61 377. Accessed April 26, 2007.

Gonsalves, C., D. Lee, and D. Gonsalves. 2007. "The Adoption of Genetically Modified Papaya in Hawaii and Its Implications for Developing Countries." *Journal of Development Studies* 43(1): 177–191.

Goodman, D. 1991. "Some Recent Tendencies in the Industrial Reorganization of the Agri-food System." *Towards a New Political Economy of Agriculture*. W. Friedland, L. Busch, F. Buttel, and A. Rudy, eds. Westview Special Studies in Agriculture Science and Policy. Boulder: Westview Press. 37–64.

Goodman, D., B. Sorj, and J. Wilkinson. 1987. *From Farming to Biotechnology: A Theory of Agro-Industrial Development*. Oxford: Basil Blackwell.

Goodman, D., and M. Watts. 1994. "Reconfiguring the Rural or Fording the Divide? Capitalist Restructuring and the Global Agro-Food System." *Journal of Peasant Studies* 22(1): 1–49.

Gordon, D., R. Edwards, and M. Reich. 1994. "Long Swings and Stages of Capitalism." *Social Structures of Accumulation*. Cambridge: Cambridge University Press. 11–28.

Gouveia, L. 1997. "Reopening Totalities: Venezuela's Restructuring and the Globalization Debate." *Globalising Food: Agrarian Questions and Global Restructuring*. D. Goodman and M. Watts, eds. London: Routledge. 305–323.

Grace, B. 2002. "Patents and Plants: How Developing Countries Are Protecting Their Genetic Resources." *Living in Hope: People Challenging Globalization*. J. Feffer, ed. London: Zed Books. 84–96.

Greenbaum, A., and A. Wellington. 2002. "Introduction: A 'Law and Society' Approach to Environmental Law." *Environmental Law in Social Context: A Canadian Perspective*. A. Greenbaum, A. Wellington, and R. Pushchak, eds. Concord: Captus Press. 2–21.

Guigni, M. 2008. "Political, Biographical, and Cultural Consequences of Social Movements." *Sociology Compass* 2(5): 1582–1600.

Hamilton, N. D. 2005. "Forced Feeding: New Legal Issues in the Biotechnology Policy Debate." *Washington University Journal of Law and Policy* 17: 37–58.

Hansen, A. B. 2004. "Mrs. Schmeiser Sues Monsanto for $140." Common Ground. Available online: http://www.commonground.ca/iss/0412161/cg161_MrsSchmeiser. Accessed November 17, 2006.

Harvey, D. 2003. *The New Imperialism*. Oxford: Oxford University Press.

Hendrickson, M., and W. Heffernan. 2007. *Concentration of Agricultural Markets*. Available online: http://civileats.com/wp-content/uploads/2009/05/2007-heffernanreport.pdf. Accessed September 20, 2009.

Holzer, B. 2001. "Transnational Protest and the Corporate Planet: The Case of Mitsubishi Corporation vs. The Rainforest Action Network." *Asian Journal of Social Science* 29(1): 73–86.

International Society for Horticultural Science. Horticulture Research International. "Canada—Climate." Available online: http://www.hridir.org/countries/canada/index.htm. Accessed July 12, 2006.

James, C. 2000. "Global Status of Commercialized Transgenic Crops—2000." International Service for the Acquisition of Agricultural Biotechnology Applications. ISAAA Brief 21. Available online: http://www.isaaa.org/resources/publications/briefs/21/default.html. Accessed July 15, 2011.

———. 2008. "Global Status of Commercialized Biotech/GM Crops: 2008." International Service for the Acquisition of Agricultural Biotechnology Applications. ISAAA Brief 39: Executive Summary. Slides and Tables. Available online: http://www.isaaa.org/resources/publications/briefs/39/executivesummary/default.html. Accessed May 15, 2009.

———. 2010. "Global Status of Commercialized Biotech/GM Crops: 2010." International Service for the Acquisition of Agricultural Biotechnology Applications. ISAAA Brief 42: Executive Summary. Available online: http://www

.isaaa.org/resources/publications/briefs/42/executivesummary/default.asp. Accessed: June 28, 2011.

Jasanoff, Sheila. 1995. "Legal Encounters with Genetic Engineering." *Science at the Bar: Law, Science, and Technology in America*. Cambridge: Harvard University Press. 138–159.

Jefferson, A., C. Correa, G. Otero, D. Blyth, and C. Qualset. 1999. "Genetic Use Restriction Technologies: Technical Assessment of the Set of New Technologies Which Sterilize or Reduce the Agronomic Value of Second Generation Seed, as Exemplified by U.S. Patent 5,723,765, and WO 94/03619." Montreal: United Nations Convention on Biological Diversity. Available online: http://www.biodiv.org/sbstta4/HTML/SBSTTA4-9-rev1e.html. Accessed September 9, 2008.

Jones, K. E. 2000. "Constructing rBST in Canada: Biotechnology, Instability and the Management of Nature." *Canadian Journal of Sociology* 25(3): 311–341.

Kaiser, C. 2005. "Take It or Leave It: Monsanto v. McFarling, Bowers v. Baystate Technologies, and the Federal Circuit's Formalistic Approach to Contracts of Adhesion." *Chicago-Kent Law Review* 80: 487–514.

Kautsky, K. 1988 [1899]. *The Agrarian Question*. Vol. 1. Trans. P. Burgess. London: Zwann Publications.

Kershen. 2004. "Of Straying Crops and Patent Rights." *Washburn Law Journal* 43: 575–610.

Kilman, S. 2010. "Monsanto Draws Antitrust Scrutiny." CheckBiotech.org. Available online: http://greenbio.checkbiotech.org/news/monsanto_draws_antitrust_scrutiny. Accessed November 15, 2011.

Kinchy, A., D. Kleinman, and R. Autry. 2008. "Against Free Markets, against Science? Regulating the Socio-Economic Effects of Biotechnology." *Rural Sociology* 73(2): 147–179.

Kloppenburg, J. 2004. *First the Seed: Political Economy of Plant Biotechnology, 1492–2000*. 2nd ed. Madison: University of Wisconsin Press.

———. 2010. Impeding Dispossession, Enabling Repossession: Biological Open Source and the Recovery of Seed Sovereignty. *Journal of Agrarian Change* 10(3): 367–388.

Kotz, D. M. 1994. "Interpreting the Social Structure of Accumulation Theory." *Social Structures of Accumulation*. Cambridge: Cambridge University Press. 50–71.

Kurzer, P., and A. Cooper. 2007. "What's for Dinner? European Farming and Food Traditions Confront American Biotechnology." *Comparative Political Studies* 40: 1035–1058.

Kuyek, D. 2005. *Reaping What's Sown: How the Privatization of the Seed System Will Shape the Future of Canadian Agriculture*. MA Thesis. University of Quebec at Montreal. Available online: http://www.forumonpublicdomain.ca/node/48 Accessed August 10, 2006.

Lassen J., and P. Sandoe. 2009. "GM Farmers and the Public—A Harmonious Relation." *Sociologia Ruralis* 49(3): 258–272.

Le Heron, R., and M. Roche. 1995. "A 'Fresh' Place in Food's Space." *Area* 27(1): 23–33.

Leiss, W., and M. Tyshenko. 2001. "Some Aspects of the 'New Biotechnology' and Its Regulation in Canada." *Canadian Environmental Policy*. 2nd ed. D. Van Nijnatten and R. Boardman, eds. Oxford: Oxford University Press. 321–343.

Lenin, V. I. 1964 [1899]. *The Development of Capitalism in Russia*. 2nd ed., rev. Moscow: Progress Publishers.

Lewontin, R. C. 2000. "The Maturing of Capitalist Agriculture: Farmer as Proletarian." *Hungry for Profit*. 93–106.

Lies, M. 2011. "Coba Presses Scotts for Bentrgrass Plan." *Capital Press*. Available online: http://www.capitalpress.com/oregon/ml-coba-letter-021111. Accessed July 6, 2011.

Lyons, M. 2001. "Monsanto Ready to Wage War." *Saskatoon Star Phoenix*. July 19, 2001. Available online: http://www.percyschmeiser.com/war.htm. Accessed November 24, 2006.

Mandel, G. 2004. "Gaps, Inexperience, Inconsistencies, and Overlaps: Crisis in the Regulation of Genetically Modified Plants and Animals." *William and Mary Law Review*. Available online: http://www.highbeam.com/. Accessed: January 10, 2005.

Mandel, G. 2005. "The Future of Biotechnology Litigation and Adjudication." *Pace Environmental Law Review* 23: 83–112.

Marchant, Gary. 1988. "Modified Rules for Modified Bugs: Balancing Safety and Efficiency in the Regulation of Deliberate Release of Genetically Engineered Microorganisms." *Harvard Journal of Law and Technology* 1: 163–208. Available online: http://jolt.law.harvard.edu/articles/pdf/01HarvJLTech163.pdf. Accessed July 11, 2007.

Mascarenhas, M., and L. Busch. 2006. "Seeds of Change: Intellectual Property Rights, Genetically Modified Soybeans and Seed Saving in the United States." *Sociologia Ruralis* 46(2): 122–138.

Mauro, I., and S. McLachlan. 2008. "Farmer Knowledge and Risk Analysis: Post-release Evaluation of Herbicide-Tolerant Canola in Western Canada." *Risk Analysis* 28: 463–476.

McAfee, K. 2003. "Biotech Battles: Plants, Power and Intellectual Property in the New Global Governance Regimes." *Globalization and Its Discontents*. R. Schurman and D. Kelso, eds. Berkeley: University of California Press. 174–194.

McBride, S. 2001. *Paradigm Shift: Globalization and the Canadian State*. Halifax: Fernwood Publishing.

McBride, S., and J. Shields. 1997. *Dismantling a Nation: The Transition to Corporate Rule in Canada*. Halifax: Fernwood Publishing.

McCann, Michael. 1994. *Rights at Work: Pay Equity Reform and the Politics of Legal Mobilization*. Chicago: University of Chicago Press.

———. 2004. "Law and Social Movements." *The Blackwell Companion to Law and Society*. A. Sarat, ed. Malden, MA: Blackwell. 506–522.

———. 2006. "Law and Social Movements: Contemporary Perspectives." *Annual Review of Law and Social Science* 2: 17–38.

McEowen, R. 2004. "Legal Issues Related to the Use and Ownership of Genetically Modified Organisms." *Washburn Law Journal* 43: 611–659.

McLeod-Kilmurray, H. 2007. "Hoffman v. Monsanto: Courts, Class Actions, and Perceptions of the Problem of GM Drift." *Bulletin of Science, Technology and Society* 27(3): 188–201.

McMichael, P. 1991. "Food, the State, and the World Economy." *International Journal of Sociology of Agriculture and Food* 1: 71–85.

———. 1992. "Tensions between National and International Control of the World Food Order: Contours of a New Food Regime." *Sociological Perspectives* 35(2): 343–365.

———. 1995. "Introduction." *Food and Agrarian Orders in the World-Economy*. P. McMichael, ed. Westport, CT: Greenwood Press.

———. 1996. "Globalisation: Myths and Realities." *Rural Sociology* 61(1): 25–55.

———. 2004. "Global Development and the Corporate Food Regime." Paper prepared for the Symposium on New Directions in the Sociology of Global Development. Eleventh World Congress of Rural Sociology. Trondheim. July 2004.

———. 2005. "Global Development and the Corporate Food Regime." *Research in Rural Sociology and Development* 11: 269–303. Available online: http://www.agribusinessaccountability.org/pdfs/297_Global%20Development%20and%20the%20Corporate%20Food%20Regime.pdf. Accessed January 18, 2006.

———. 2009. "The Agrofuels Project at Large." *Critical Sociology* 35(6): 825–839.

McNally, R., and P. Wheale. 1998. "The Consequences of Modern Genetic Engineering: Patents, 'Nomads' and the 'Bio-Industrial Complex.'" *The Social Management of Genetic Engineering*. P. Wheale, R. von Schomberg, and P. Glasner, eds. London: Ashgate. 303–330.

Meyer, L. 2011. "Genetically Modified Bentgrass Continues to Spread in County. Argus Observer. Available online: http://www.argusobserver.com/articles/2011/02/27/news/doc4d6a0e854ed35657251680.txt. Accessed July 6, 2011.

Middendorf, G., M. Skladney, E. Ransom, and L. Busch. 2000. "New Agricultural Biotechnologies: The Struggle for Democratic Choice." *Hungry for Profit*. New York: Monthly Review Press. 93–106.

Minder, Raphael. 2006. "GM Foods Verdict Unlikely to Alter EU Rules." FT.Com *Financial Times*. January 4, 2006. Available online: http://news.ft.com/cms/s/c39f43ae-7d5b-11da-875c-0000779e2340.html. Accessed January 4, 2006.

Mitcham, C. 1995. "The Concept of Sustainable Development: Its Origins and Ambivalence." *Technology in Society* 17(3): 311–326.

Mittelstaedt, M. 2009. "Attack of the Triffids Has Flax Farmers Baffled." *Globe and Mail*. Available online: http://gefreebc.wordpress.com/2009/11/13/monsanto-triffid-flax-canada/. Accessed July 6, 2011.

Moeller, D., and M. Sligh. 2004. *Farmers' Guide to GMOs*. K. Krub, ed. St Paul, MN: Farmers' Legal Action Group, Inc., and Rural Advancement Foundation International.

Monsanto Company. 2005. "Putting Technology to Work in the Field." Monsanto Company, promotional flyer. January 1, 2005.

———. 2006. 2006 Monsanto Technology/Stewardship Agreement. (Limited Use Licence). St. Louis, MO: The Monsanto Company. Copy with author.

Monsanto vs. Schmeiser. Website. Available online: http://www.percyschmeiser.com/. Accessed September 5, 2006.

Moran, W., G. Blunden, M. Workman, and A. Bradly. 1996. "Family Farmers,

Real Regulation, and the Experience of Food Regimes." *Journal of Rural Studies* 12(3): 245–258.

Mudeva, A. 2004. "Controversial Dutch Co-existence Rules—Dutch Farmers Agree on GMO Crop Separation Rules." Reuters. November 2, 2004. Available online: http://www.grain.org/research/contamination.cfm?id=225. Accessed December 11, 2006.

Müller, B. 2008. "Still Feeding the World? The Political Ecology of Canadian Prairie Farmers." *Anthropologica* 50(2): 389–407.

Murphy, R. 1994. *Rationality and Nature: A Sociological Inquiry into a Changing Relationship*. Boulder: Westview Press.

Nagatada, Takayanagi. 2006. "Global Flows of Fruit and Vegetables in the Third Food Regime." *Journal of Rural Community Studies* 102: 25–41.

NelsonFarm.Net: Information on Nelson vs. Monsanto. Available online at http://nelsonfarm.net/. Accessed October 5, 2006.

Netstate.com. "Mississippi: The Geography of Mississippi." Netstate.com Website. Available online: http://www.netstate.com/states/geography/ms_geography.htm. Accessed May 10, 2006.

Neuman, W., and A. Pollack. 2010. "Farmers Cope with Roundup Resistent Weeds." *New York Times*. Available online: http://www.nytimes.com/2010/05/04/business/energy-environment/04weed.html. Accessed August 24, 2011.

"New Study Indicates Innovation Place Has Significant Economic Impact." 1996. Available online: http://www.innovationplace.com/html/newslttr/1996/Apr.1996/b1.html. Accessed July 25, 2006.

Nottenburg, C., and J. Sharples. "Key Organisations and Agreements in IP Tutorial." Biological Innovation for Open Society Website. Available online: http://www.bios.net/daisy/KeyOrgs/1236/1337.html. Accessed December 8, 2006.

Novek, J. 2003. "Intensive Hog Farming in Manitoba: Transnational Treadmills and Local Conflicts." *Canadian Review of Sociology and Anthropology* 40(1): 3–26.

Ó Riain, S. 2000. "States and Markets in an Era of Globalization." *Annual Review of Sociology* 26: 187–213.

Otero, G. 1995a. "Agricultural Biotechnology in Latin America: Studying Its Future Impacts." International Development Research Center. Available online: http://www.irdc.ca/books/focus/789/otero.html. 1–11.

———. 1995b. "The Coming Revolution of Biotechnology: A Critique of Buttel." *The Biotechnology Revolution?* M. Fransman, G. Junne, and A. Roobeek, eds. Oxford: Blackwell.

Otero, G., and G. Pechlaner. 2005. "Food for the Few: The Biotechnology Revolution in Latin America." *Canadian Journal of Development Studies* 26: 867–887.

———. 2008. "Latin American Agriculture, Food and Biotechnology: Temperate Dietary Pattern Adoption and Unsustainability." *Food for the Few: Neo-Liberal Globalism and the Biotechnology Revolution in Latin America*. G. Otero, ed. Austin: University of Texas Press.

Park, S. 2003. "Battle of Titans: Intellectual Property Regime v. UCC." *University of Illinois Journal of Law, Technology and Policy* 2: 531–549.

Pechlaner, G. 2010. "Biotech on the Farm: Mississippi Agriculture in an Age of Proprietary Biotechnologies." *Anthropologica* 52: 291–304.

————. Undated. "GMO-Free America, or Just Drink Your Organic Wine and Be Happy? Mendocino County and the Impact of Local Level Resistance to the Agricultural Biotechnology Paradigm." Unpublished paper.

Pechlaner, G., and G. Otero. 2010. "Neoliberal Globalism and the Third Food Regime: Neoregulation and the New Division of Labor in North America." *Rural Sociology* 75(2): 179–208.

Pew Initiative on Food and Biotechnology. 2001. "Guide to U.S. Regulation of Genetically Modified Food and Agricultural Biotechnology Products." Available online: http://pewagbiotech.org/resources/issuebriefs/1-regguide.pdf. Accessed December 5, 2005.

————. 2003. "2001–2002 Legislative Activity Related to Agricultural Biotechnology." Available online: http://pewagbiotech.org/ Accessed January 5, 2005.

————. 2004. "Genetically Modified Crops in the United States." Available online: http://pewagbiotech.org/resources/factsheets/display.php3?FactsheetID=2. Accessed January, 11, 2005.

————. 2005. "U.S. vs E.U.: An Examination of the Trade Issues Surrounding Genetically Modified Food." Available online: http://pewagbiotech.org/re sources/issuebriefs/useu.pdf. Accessed January 4, 2005.

Pfeffer, Max. 1992. "Sustainable Agriculture in Historical perspective." *Agriculture and Human Values* 9(4): 4–11.

Polanyi, K. [1944] 2001. *The Great Transformation*. New York and Toronto: Rinehart and Co.

Pratt, S. 2005. "Roundup Ready Canola Back in Schmeiser's Field." *The Western Producer*. October 26, 2005. Available online: http://www.percyschmeiser.com/ RRCanola%20Returns.htm. Accessed November 24, 2006.

Reuters. 2005. "Lawmaking on Genetic Food is Minefield for EU." Available online: http://www.healthypages.net/newspage.asp?newsid=4920. February 28, 2005. Accessed January 10, 2006.

Roberts, T. E. 1999. "Life Forms as Patentable Subject Matter: Is a Divergence in Canadian and U.S. Laws Warranted?" Saskatoon, Saskatchewan: Benesh Bitz and Company. 1–42.

Royal Society of Canada. 2001. "Elements of Precaution: Recommendations for the Regulation of Food Biotechnology in Canada." An expert panel report prepared at the request of Health Canada, the Canadian Food Inspection Agency, and Environment Canada. Available online: http://www.rsc.ca/index .php?page=expert_panels_food&lang_id=1&page_id=119. Accessed October 18, 2005.

Sanderson, S. E. 1986. "The Emergence of the "World Steer": Internationalization and Foreign Domination in Latin American Cattle Production." *Food, the State and International Political Economy: Dilemmas of Developing Countries* F. L. Tullis and W. L. Hollist, eds. Lincoln: University of Nebraska Press.

Schnaiberg, A., and K. A. Gould. 1994. *Environment and Society: The Enduring Conflict*. New York: St. Martin's Press.

Schubert, R. 2001. "Monsanto Still Suing Nelson, Other Growers." *CropChoice News*. May 21, 2001. Available online: http://nelsonfarm.net/issue.htm. Accessed October 5, 2006.

Schurman, R., and D. Kelso (Eds.). 2003. *Engineering Trouble: Biotechnology and Its Discontents*. Berkeley: University of California Press.

Scott, K. 2010. "Monsanto Draws Antitrust Scrutiny." *Wall Street Journal*. Available online: http://online.wsj.com/article/SB1000142405274870370100457511 3911550788020.html. Accessed August 23, 2011.

Seifert, F. 2008. "Consensual NIMBYs, Contentious NIABYs: Explaining Contrasting Forms of Farmers GMO Opposition in Austria and France." *Sociolgia Ruralis* 49(1): 20–40.

Shiva, V. 2000a. "North-South Conflicts in Intellectual Property Rights." *Peace Review* 12(4): 501–508.

———. 2000b. *Stolen Harvest: The Hijacking of the Global Food Supply*. Cambridge: South End Press.

———. 2001. *Protect or Plunder? Understanding Intellectual Property Rights*. London: Zed Books.

Strange, S. 1996. *The Retreat of the State: The Diffusion of Power in the World Economy*. New York: Cambridge University Press.

———. "The Declining Authority of States." *The Global Transformations Reader: An Introduction to the Globalization Debate*. D. Held and A. McGrew, eds. Cambridge: Polity Press. 127–134.

Suchitra, M., and C. Surendaranath. 2004. Kerala Farmers Grow Vanilla for Profit. *Down to Earth*. Available online: http://www.indiaenvironmentportal.org.in/node/37401. Accessed September 20, 2009.

Sumner, C. 1979. "The Ideological Composition of Law." *Reading Ideologies: An Investigation into the Marxist Theory of Ideology and Law*. London: Academic Press.

Teeple, Gary. 2000. *Globalization and the Decline of Social Reform*. 2nd ed. Toronto: Garamond Press.

Thompson, S., and J. T. Cowan. 1995. "Durable Food Production and Consumption in the World-Economy." *Food and Agrarian Orders in the World-Economy*. P. McMichael, ed. Westport, CT: Greenwood Press. 35–54.

Tokar, B. (Ed.). 2001. *Redesigning Life?* New York: Zed Books.

Union of Concerned Scientists. Undated. "Rice 'Mystery' Illustrates Potential for Food Contamination with Unapproved, Genetically Altered Crops." Available online: http://www.ucsusa.org/food_and_agriculture/science_and_impacts/impacts_genetic_engineering/rice-contamination-a-mystery.html. Accessed October 14, 2009.

University of Saskatchewan. 2004. "Organic Statistics 2003. Saskatchewan." University of Saskatchewan, Organic Information Website. From "'Certified Organic': The Status of the Canadian Organic Market in 2003." Prepared for Agriculture and Agri-food Canada by Anne Macey, March 2004. Available online: http://organic.usask.ca/statistics.htm. Accessed January 2005.

Urmetzer, P. 2005. *Globalization Unplugged*. Toronto: University of Toronto Press.

Vaver, D. 2004. "Canada's Intellectual Property Framework: A Comparative Overview." *Intellectual Property Journal* 17: 125–188.

Vidal, John. 1999. "World Braced for Terminator 2." *The Guardian*. Available online: http://www.guardian.co.uk/science/1999/oct/06/gm.food2. Wednesday, October 6, 1999. Accessed November 23, 2009.

Voosen, P. 2010. "Judge Orders Destruction of Genetically Modified Beets."

New York Times. Available online: http://www.nytimes.com/gwire/2010/12/ 01/01greenwire-judge-orders-destruction-of-genetically-modifi-66587.html. Accessed July 18, 2011.

Walsh-Dilley, M. 2009. Localizing Control: Mendocino County and the Ban on GMOs. *Agriculture and Human Values* 26: 95–105.

Warick, Jason. 2001. "GM Flax Seed Yanked off Canadian Market— Rounded Up, Crushed." *Star Phoenix*. June 23, 2001. Available online: http://www.organic consumers.org/patent/flaxrecall.cfm. Accessed August 2, 2006.

Weiss, L. 1997. "Globalization and the Myth of the Powerless State." *New Left Review* 225: 3–27.

Wells, M. 1997. "Legal Discourse and the Restructuring of Californian Agriculture: Class Relations at the Local Level." *Globalising Food: Agrarian Questions and Global Restructuring*. D. Goodman and M. Watts, eds. London: Routledge. 235–255.

Wells, Stuart, and Holly Penfound. 2003. "Canadian Wheat Board Speaks out against Roundup Ready Wheat." *Toronto Star*, Ontario Edition. February 25, 2003. Available online: http://www.organicconsumers.org/ge/031803_ge_ wheat.cfm. Accessed July 28, 2006.

Whittier, N. 2004. "The Consequences of Social Movements for Each Other." *The Blackwell Companion of Social Movements*, D. Snow, S. A. Soule, and H. Kriesi, eds. Malden, MA: Blackwell. 531–552.

Wield, D., Chataway, J., and Bolo, M. 2010. "Issues in the Political Economy of Agricultural Biotechnology." *Journal of Agrarian Change* 10(3): 342–366.

Witte, B. 2001. "Board Recommends Dismissal of GMO Suit." *AgWeek*. April 23, 2001. Available online: http://nelsonfarm.net/. Accessed October 5, 2006.

Wolfson, M. 2003. "Neoliberalism and the Social Structure of Accumulation." *Review of Radical Political Economics* 35(3): 255–262.

Wright, Tom. 2005. "U.S. Fines Swiss Company over Sale of Altered Seed." *New York Times*. Available online: http://www.nytimes.com/2005/04/09/business/ worldbusiness/09syngenta.html. Accessed April 9, 2005.

Yin, R. K. 1982. "Studying Phenomenon and Context across Sites." *American Behavioral Scientist* 26(1): 84–100.

———. 2003. *Case Study Research: Design and Methods*. 3rd ed. Applied Social Research Methods Series. Thousand Oaks, CA: Sage Publications.

Government and Institutional Documents

Canadian Biotechnology Advisory Committee. 2002. "Improving the Regulation of Genetically Modified Foods and Other Novel Foods in Canada." CBAC. Available online: http://cbac-cccb.ca/epic/internet/incbac-cccb.nsf/en/Home. Accessed October 18, 2005.

———. 2004. "Completing the Biotechnology Regulatory Framework." February 2004 Advisory Memorandum. CBAC. Available online: http://cbac-cccb .ca/epic/site/cbac-cccb.nsf/en/ah00436e.html. Accessed October 18, 2005.

Canadian Intellectual Property Office. "A Guide to Patents: Patent Protection."

Available online: http://strategis.ic.gc.ca/sc_mrksv/cipo/patents/pat_gd_pro tect-e.html#sec2. Accessed: September 27, 2006.

———. Canadian Intellectual Property Office Website. Available online: http:// strategis.ic.gc.ca/sc_mrksv/cipo/patents/pat_gd_protect-e.html#sec1. Accessed September 6, 2006.

Canadian Manual of Patent Office Practice. 2004 [1998]. Available online: http:// strategis.ic.gc.ca.proxy.lib.sfu.ca/sc_mrksv/cip/patents/mopop/mopope.pdf. Accessed September 16, 2006.

Codex Alimentarius Website. Available online: http://www.codexalimentarius .net/web/index_en.jsp. Accessed December 8, 2006.

Commission of the European Communities. 2002. "Communication from the Commission to the Council, the European Parliament, the Economic and Social Committee and the Committee of the Regions. Life Sciences and Biotechnology—A Strategy for Europe." Available online: http://eur-lex.europa .eu/LexUriServ/site/en/com/2002/com2002_0027en01.pdf. Brussels. Accessed January 23, 2002.

Convention on Biological Diversity [CBD]. "Background." CBD Website. Available online: http://www.biodiv.org/doc/publications/guide.shtml?id=web. Accessed December 9, 2006.

———. "Sustaining Life on Earth." CBD Website. Available online: http://www .biodiv.org/doc/publications/guide.shtml?id=web. Accessed December 9, 2006.

Europa, Website of the European Union. 2005a. "Commission Authorizes Danish State Aid to Compensate for Losses due to Presence of GMOs in Conventional and Organic Crops." http://europa.eu.int/comm/food/food/biotech nology/index_en.htm. Accessed December 10, 2005.

———. 2005b. "Questions and Answers on the Regulation of GMOs in the European Union." Available online: http://europa.eu.int/comm/food/food/bio technology/index_en.htm. Accessed December 10, 2005.

———. 2005c. "Transboundary Movement of Genetically Modified Organisms." Available online: http://europa.eu.int/comm/food/food/biotechnology/ index_en.htm. Accessed December 10, 2005.

———. 2010. "GMOs: Member States to Be Given Full Responsibility on Cultivation in Their Territories." Available online: http://europa.eu/rapid/ pressReleasesAction.do?reference=IP/10/921. Accessed July 5, 2011.

Food and Agriculture Organization [FAO], Commission on Genetic Resources for Food and Agriculture. "International Treaty on Plant Genetic Resources for Food and Agriculture" [ITPGR]. Available online: http://www.fao.org/ AG/cgrfa/itpgr.htm. Accessed December 8, 2006.

———. "FAO Statement on Biotechnology." Available online: http://www.fao .org/Biotech/stat.asp. Accessed September 5, 2006.

Government of Canada. Canadian Food Inspection Agency. "Proposed Amendments to the Plant Breeders' Rights Act to Bring Existing Legislation into Conformity with the 1991 UPOV Convention." CFIA Website. Available online: http://www.inspection.gc.ca/english/plaveg/pbrpov/ammende.shtml. Accessed October 10, 2006.

———. Science Branch. Office of Biotechnology. "Federal Government Agrees on New Regulatory Framework for Biotechnology." Available online: http://www.inspection.gc.ca/english/sci/biotech/reg/fracade.shtml. Accessed December 12, 2005.

Government of Canada, Canadian Food Inspection Agency, Health Canada, and USDA. 1998. "Canada and United States Bilateral on Biotechnology." Available online: http://www.aphis.usda.gov/brs/canadian/usda01e.pdf. Accessed: November 20, 2011.

Government of Canada. Canadian General Standards Board. 2004. Voluntary Labeling and Advertising of Foods That Are and Are not Products of Genetic Engineering. CAN/CGSB-32.315-2004 ICS 55.020. Available online: www.ongc-cgsb.gc.ca.

Government of Canada. Industry Canada. 1998. "The 1998 Canadian Biotechnology Strategy: An Ongoing Renewal Process." Available online: http://www.biostrategy.gc.ca/CMFiles/1998strategyE49RAI-8312004-5365.pdf. Accessed November 15, 2011.

Government of Saskatchewan, Bureau of Statistics. 2004. Saskatchewan Fact Sheet. Available online: http://www.stats.gov.sk.ca/docs/factsheet05.pdf. Accessed May 10, 2006.

Government of Saskatchewan. "About Saskatchewan: Facts and Figures." Available online: http://www.gov.sk.ca/aboutsask/. Accessed July 11, 2006.

———. 2005. "Research Parks Worth Half a Billion Dollars to Province." News Release. February 17, 2005. Available online: http://www.innovationplace.com/html/frameset.html. Accessed July 25, 2006.

International Union for the Conservation of Nature and Natural Resources/World Conservation Union. "Article 9—Farmers' Rights." Available online: http://www.iucn.org/bookstore/HTML-books/EPLP057-expguide-international-treaty/Article9.html. Accessed October 10, 2006.

International Union for the Protection of New Varieties of Plants. 1978. International Convention for the Protection of New Varieties of Plants of December 2, 1961, as Revised at Geneva on November 10, 1972, and on October 23, 1978. Available online: http://www.upov.int/en/publications/conventions/1978/act1978.htm. Accessed October 10, 2006.

———. 1991. International Convention for the Protection of New Varieties of Plants of December 2, 1961, as Revised at Geneva on November 10, 1972, on October 23, 1978, and on March 19, 1991. Available online: http://www.upov.int/en/publications/conventions/1991act1991.htm. Accessed September 25, 2006.

Mississippi Department of Agriculture and Commerce. "Mississippi Agriculture at a Glance." Available online: http://www.mdac.state.ms.us/n_library/misc/ag_overview.html. Accessed May 3, 2006.

Mississippi State University Extension Service Website. "Crops: Cotton." MSU Cares, Coordinated Access to the Research and Extension System Website. Available online: http://msucares.com/crops/cotton/index.html. Accessed May 3, 2005.

National Research Council Canada. 2002. "Saskatoon's Biotechnology Cluster." *PBI Bulletin.* 2002 Issue 3. Available online: http://pbi-ibp.nrc-cnrc.gc.ca/en/bulletin/2002issue3/page5.htm. Accessed July 10, 2006.

———. 2005. "Saskatoon Agricultural Biotechnology, Nutraceuticals, and Bio-Products." Available online: http://www.nrc-cnrc.gc.ca/clusters/saskatoon_e .html. Accessed July 20, 2006.

Official Journal of the European Communities. 2001. "Directive 2001/18/EC of the European Parliament and of the Council of 12 March 2001. On the Deliberate Release into the Environment of Genetically Modified Organisms and Repealing Council Directive 90/220/EEC." Available online: http://www.bio safety.be/PDF/2001_18.pdf. Accessed November 22, 2011.

———. 2003. "Concerning the Traceability and Labeling of Genetically Modified Organisms and the Traceability of Food and Feed Products Produced from Genetically Modified Organisms and Amending Directive 2001/18/EC." Regulation (EC) No. 1830/2003 of the European Parliament and of the Council of September 22, 2003.

Saskatchewan Agriculture and Food. 2007. "'Number of Census Farms' and 'Size of Farm.'" Census StatFact. May 16, 2007. Available online: http://www .agriculture.gov.sk.ca/Default.aspx?DN=bcd1ff12-9da2-4023-b941-4137b4 ca8ee9. Accessed December 6, 2009.

Saskatchewan Agriculture, Food and Rural Revitalization. 2002a. "Saskatchewan Farm Land Use." Census StatFact. 2001 Census. May 15, 2002. Available online: http://www.agr.gov.sk.ca/docs/statistics/finance/other/SkFarmLandUse .pdf. Accessed May 10, 2006.

———. 2002b. "Type of Farm." Census StatFact. 2001 Census. May 15, 2002. Available online: http://www.agr.gov.sk.ca/docs/statistics/finance/other/Type Farm.pdf. Accessed May 10, 2006.

Statistics Canada. 2001. "Sharp Decline in Number of Farms in Saskatchewan." 2001 Census of Agriculture. Available online: http://www.statcan.ca/english/ agcensus2001/first/regions/farmsk.htm. Accessed July 11, 2006.

———. 2004. "Selected Oilseeds, by Provinces (1981–2001 Censuses of Agriculture)." Available online: http://statcan.ca/english/Pgdb/agrc29i.html. Accessed November 8, 2004.

United States Code. Updated March 2000. "United States Code Title 35—Patent Laws." Available online at Bitlaw: http://www.bitlaw.com/source/35usc/. Accessed October 2, 2006.

United States Congress, Office of Technology Assessment. 1989. New Developments in Biotechnology: Patenting Life—Special Report. OTA-BA-370. Washington, DC: U.S. Government Printing Office. Available online: http://govinfo .library.unt.edu/ota/Ota_2/DATA/1989/8924.PDF#search=%22New%20 Developments%20in%20Biotechnology%3A%20Patenting%20Life% E2%80%94Special%20Report%22. Accessed October 2, 2006.

United States Department of Agriculture, Economic Research Service, 2009. "Adoption of Genetically Engineered Crops in the U.S." Available online: http://www.ers.usda.gov/Data/BiotechCrops/. Accessed: November 27, 2011.

———. "State Fact Sheets: Mississippi." Available online: http://www.ers.usda .gov/StateFacts/MS.HTM. Accessed May 20, 2006.

United States Department of Agriculture, National Agricultural Statistics Service. 2005. "Acreage." Available online: http://usda.mannlib.cornell.edu/re ports/nassr/field/pcp-bba/acrg0605.pdf. Accessed May 23, 2006.

————. "Mississippi State Agriculture Overview-2005." Available online: http://www.nass.usda.gov/Statistics_by_State/Ag_Overview/AgOverview_MS.pdf. Accessed May 3, 2005.

————. "Table 41: Farms by Concentration of Market Value of Agricultural Products Sold: 2002." 2002 Census of Agriculture. Vol. 1, Chap. 1, Mississippi State Level Data. Available online: http://www.nass.usda.gov/census/census02/volume1/ms/index1.htm. Accessed April 27, 2007.

United States Department of Agriculture, Office of the Inspector General, Southwest Region. 2005. *Audit Report: Animal and Plant Health Inspection Service Controls over Issuance of Genetically Engineered Organism Release Permits*. Audit 50601-Te. December 2005. Available online: http://nsrg.neu.edu/resources/regulatory_capacity/documents/5060108TE.pdf.

United States. *Federal Register*. 2009. "Agriculture and Antitrust Enforcement Issues in Our Twenty-First Century Economy." August 27, 2009. *Federal Register* 74(165). Available online: http://edocket.access.gpo.gov/2009/pdf/E9-20671.pdf. Accessed December 6, 2009.

United States. *Federal Register*. Office of Science and Technology Policy. 1986. "Coordinated Framework for Regulation of Biotechnology: Announcement of Policy and Notice for Public Comment." June 26, 1986. *Federal Register* 51(123):23302–23350.

World Trade Organization. "Legal Texts: The WTO Agreements." Available online: http://www.wto.org/english/docs_e/legal_e/ursum_e.htm#nAgreement. Accessed December 8, 2006.

————. "Trade Related Aspects of Intellectual Property." Available online: http://www.wto.org/english/docs_e/legal_e/27-trips_04c_e.htm. Accessed December 8, 2006.

Index

See page vii for a complete list of acronyms and their full titles.

Lightning Source UK Ltd.
Milton Keynes UK
UKHW021600010320
359497UK00011B/211